Modernizing Legacy Applications to Microsoft Azure

Plan and execute your modernization journey seamlessly

Steve Read

Larry Mead

BIRMINGHAM—MUMBAI

Modernizing Legacy Applications to Microsoft Azure

Copyright © 2023 Packt Publishing

Group Product Manager: Preet Ahuja
Publishing Product Manager: Preet Ahuja
Senior Editor: Runcil Rebello
Technical Editor: Irfa Ansari
Copy Editor: Safis Editing
Project Coordinator: Ashwin Kharwa
Proofreader: Safis Editing
Indexer: Hemangini Bari
Production Designer: Jyoti Kadam
Marketing Coordinator: Rohan Dobhal

First published: September 2023

Production reference: 1180823

Published by Packt Publishing Ltd.
Grosvenor House
11 St Paul's Square
Birmingham
B3 1RB

ISBN 978-1-80461-665-9

www.packtpub.com

I would like to acknowledge all the people who helped put this book together, especially Bob Ellsworth, who provided the foreword, and the excellent technical reviewers.

– Steve Read

To my grandchildren, for giving me inspiration for the future as we move from legacy systems to the new era of technology.

– Larry Mead

Foreword

Customers have been modernizing legacy applications for over 50 years. The pace of modernization has accelerated due to an increasing focus on business drivers, improvements in modernization technology and services, and new cloud deployment options, such as Azure.

Each customer is faced with one or more drivers to migrate, modernize, or transform. These drivers include increasing costs, decreasing skills, lack of deployment options, and lack of agility to be responsive to changing business requirements. As these drivers increase, more and more pressure is placed on **Information Technology (IT)** organizations to modernize their legacy applications.

With the increase in customer demand, migration tool vendors have invested in their technology to address this demand. This includes improving their tools to increase the level of automation and reduce manual intervention. Testing tools have also improved to accelerate the testing effort. These improvements result in lower costs, quicker migrations, and lower risk.

The cloud has been a positive revolutionary disruptor for IT. The ability to quickly deploy workloads at lower costs has accelerated the availability and adoption of new solutions to support businesses. What started as the ability to deploy virtual machines in the cloud expanded to include business and sales tools, such as Microsoft 365, **Customer Relationship Management (CRM)**, and **Enterprise Resource Planning (ERP)**. Cloud credibility has increased, ensuring that customers' most critical business applications can be cost-effectively deployed with low risk and at scale. New solutions, such as OpenAI, will continue to increase the number of customers who have embraced a cloud-first strategy. This strategy will increasingly be applied to legacy systems.

Depending on your current legacy system, a modernization project can be complex and risky. While there have been tens of thousands of successful modernization projects, some projects have failed. To ensure a successful modernization, you must understand your current legacy system, the Azure cloud platform, and options for modernizing your current applications and systems. To facilitate this process, it is best to seek advice from experts who have decades of experience in modernizing legacy applications and systems.

The modernization knowledge and advice provided in this book provide the first step in the successful modernization of your legacy applications. Once you understand these concepts, you will be ready to engage with the right legacy transformation technology and service providers in order to accelerate your modernization journey.

Bob Ellsworth

President, Mainframe Transformation Consulting

Contributors

About the authors

Steve Read currently works at Microsoft as a Global Black Belt. In this position, he helps customers migrate their legacy workloads to Azure. He has been at Microsoft for 24 years in various technical roles, focusing on application and database development. Before joining Microsoft in 1999, he was a developer on both mainframe and UNIX platforms in the banking, healthcare, and government industries.

I would like to thank all of the people who made this book a reality. It is quite an effort to publish a book like this and it takes a village to make it happen. I would like to thank the folks at Packt, who were a pleasure to work with, Bob Ellsworth and Larry Mead, for being mentors to me, and especially the technical reviewers, whose expertise helped make this book more impactful.

Larry Mead is a principal technical program manager in Microsoft Azure Core engineering. Larry has over 45 years of experience developing mission-critical systems, starting with mainframes and mid-range systems, continuing with distributed systems, and now with cloud-based systems. Larry specializes in systems for financial services, but he also has experience with retail, manufacturing, transportation, and government systems. Currently, Larry specializes in how to develop Azure services for mission-critical systems.

I would like to thank co-author Steve Read and my colleagues from my time at Microsoft, who helped review and revise the content of this book.

About the reviewers

Harri Jaakkonen is a Microsoft Valuable Professional within the security category and targets identity and access management. He is mainly focused on Microsoft technologies but also deals with network devices, load balancing, disaster recovery, and other non-Microsoft solutions. He has 27 years of experience in these areas.

Amethyst George Solomon is a senior engineering architect with the Microsoft Azure database product engineering group and collaborates with multiple product and service owners to build customer-oriented services on Azure to run mission-critical systems. He has been working with mainframe and mid-range modernization for more than a decade and has worked with customers on architecting the workload modernization strategy with the right technology. He drives product feature specifications across Azure platform products, which helps accelerate the adoption of Azure services by unblocking and accelerating complex workload migrations such as mainframes and making customer migrations successful on Azure.

Thanks to my family and friends who have been supportive throughout this time. Their support allowed me to concentrate and spend an adequate amount of time validating and adding value to this book. I would also like to thank the team behind maintaining Microsoft's documentation, for keeping it up to date, and the product group, for keeping the information precise.

Roy de Milde has been working in the field of infrastructure and application development for over 15 years for different IT companies. He has experience and knowledge in modernizing applications to leverage the latest and greatest technology and generate more business value. For the last eight years, he has been focusing on application modernization and innovation with public cloud technology for enterprise customers.

Philip Brooks holds a B.Sc. degree in computer science from Shippensburg University, as well as a PMP certificate from the Project Management Institute. He serves as a director and secretary for the SHAPE Foundation. Philip started his 34+ year career with what became Unisys as an on-site systems mainframe support analyst for the US Army under Sperry Univac. He has been active in the IT client and system support arena since 1983. He has been an operator, installer, trainer, systems internalist, and trusted consultant, and provided program and sales support across a broad spectrum of IT services and clients during his 40 years in the IT industry. Since 2020, Mr. Brooks has been part of the Microsoft Azure Core Legacy Team.

Bhaskar Bandam is an experienced professional in technology. He has been part of Microsoft's Azure Core engineering team for over three years, assisting customers with legacy workload migration to Azure. Previously, Bhaskar worked as a federal contractor, specializing in on-premises workload migration. With a career span of nearly 25 years, Bhaskar has honed his expertise in mainframe systems. He started as a mainframe developer and progressively advanced into key positions such as DevOps and digital transformation manager. Bhaskar holds a master's degree in information systems, and an MBA, strengthening his technology foundation and managerial skills.

Bhuvi Vatsey is a seasoned modernization professional with over 14 years of industry experience across the hospitality, retail, healthcare, and telecom domains as an application developer, architect, solution lead, and program manager. She has hands-on experience in transforming mission-critical applications into cloud-first and cloud-native distributed systems. She has successfully led projects with multimillion-dollar spending on migration, enhancement, and modernization of mainframe systems with customers and partners. In her current role, she is responsible for helping enterprise customers meet/exceed their modernization objectives with the right strategy and underlying solutions on Azure.

Ricardo Galvan is a senior technical program manager on the Azure Core engineering team. Ricardo provides technical assistance in migrating mainframe and mid-range applications to Azure. Before joining Microsoft, he built software tools to automate the conversion of source code of applications and databases. Ricardo specializes in converting Cobol of IBM/Unisys mainframes and IBM i (AS/400), RPG, CL, ALGOL, and other programming languages to C# and Java, among others. He is a full stack developer covering a wide range of mainframe, mid-range, and Linux/Windows programming languages, database engines, and web frameworks. He completed his first Unisys MCP mainframe COBOL and LINC banking application migration to HP-UX back in 1994.

Table of Contents

Part 2: Architecture Options

5

An Overview of the Microsoft Azure Cloud Platform   59

6

Azure Cloud Architecture Overview for Mission-Critical Workloads   77

7

Behold the Monolith – Pros and Cons of a Monolithic Architecture   95

Part 3: Azure Deployment and Future Considerations

8

Exploring Deployment Options in Azure   107

9

Modernization Strategies and Patterns – the Big Picture   117

10

Modernizing the Monolith - Best Practices   129

11

Integration – Playing Friendly in a Cloud-Native World   143

12

Planning for the Future – Trends to Be Aware of   157

Index   165

Other Books You May Enjoy   178

Preface

Hello all. This book will provide the background, options, and project planning for moving a legacy state on mainframes, mid-range, and Enterprise UNIX to the Azure cloud. This book is intended for IT architects, program managers, and project managers (alongside business groups) to understand the scope of modernization.

Who this book is for

The book will cover topics to help legacy personnel understand services in Azure, and will also help Azure personnel understand the legacy environment that is the source of the migration.

What this book covers

Chapter 1, *Understanding Your Legacy Environment - the Modernization Journey*, looks at existing legacy environments to get a baseline on the scope of modernization.

Chapter 2, *Strategies for Modernizing IBM and Unisys Mainframes*, specifically looks at IBM and Unisys mainframes, and the options for migrating these to Azure.

Chapter 3, *Midrange to Azure*, looks at options for systems that run on IBM Power, including iSeries and AIX.

Chapter 4, *Modernizing Legacy UNIX Systems*, looks at options for Enterprise UNIX systems, such as Solaris and HP-UX.

Chapter 5, *An Overview of the Microsoft Azure Cloud Platform*, looks at the various options of the Azure Cloud platform as related to migrating legacy applications.

Chapter 6, *Azure Cloud Architecture Overview for Mission-Critical Workloads*, looks at the capabilities and approaches for creating mission-critical systems on Azure.

Chapter 7, *Behold the Monolith – Pros and Cons of a Monolithic Architecture*, looks at why many legacy applications are monoliths, and the options for moving to Azure.

Chapter 8, *Exploring Deployment Options in Azure*, looks at strategies for actually moving the legacy applications and data to Azure.

Chapter 9, *Modernization Strategies and Patterns - the Big Picture*, covers strategies for moving to Azure. This includes rehosting, refactoring, and rewriting. Additionally, we cover the big bang or phased approach.

Chapter 10, Modernizing the Monolith - Best Practices, specifically looks at the challenges of modernizing monolith applications to Azure.

Chapter 11, Integration - Playing Friendly in a Cloud-Native World, looks at best practices for running in the Azure cloud.

Chapter 12, Planning for the Future - Trends to Be Aware of, looks at trends for future Azure features, and how these trends may affect your modernization.

To get the most out of this book

To get the most out of this book, we recommend that you have a background in a specific legacy environment, such as IBM or Unisys mainframe, Midrange, or Enterprise UNIX environment. Additionally, it would be helpful to have an application or system in mind for moving to Azure, or you want to build an overall strategy for moving to Azure. Finally, it would be helpful to have a background in the **service-level agreement** (**SLA**) mission-critical systems you might want to migrate.

Conventions used

There are a number of text conventions used throughout this book.

Bold: Indicates a new term, an important word, or words that you see onscreen. For instance, words in menus or dialog boxes appear in **bold**. Here is an example: "Azure Web App for Containers is technically part of what is known collectively as **Azure App Service**."

> **Tips or important notes**
> Appear like this.

Get in touch

Feedback from our readers is always welcome.

General feedback: If you have questions about any aspect of this book, email us at customercare@ packtpub.com and mention the book title in the subject of your message.

Errata: Although we have taken every care to ensure the accuracy of our content, mistakes do happen. If you have found a mistake in this book, we would be grateful if you would report this to us. Please visit www.packtpub.com/support/errata and fill in the form.

Piracy: If you come across any illegal copies of our works in any form on the internet, we would be grateful if you would provide us with the location address or website name. Please contact us at copyright@packtpub.com with a link to the material.

If you are interested in becoming an author: If there is a topic that you have expertise in and you are interested in either writing or contributing to a book, please visit authors.packtpub.com.

Share Your Thoughts

Once you've read *Modernizing Legacy Applications to Microsoft Azure*, we'd love to hear your thoughts! Scan the QR code below to go straight to the Amazon review page for this book and share your feedback.

https://packt.link/r/1804616656

Your review is important to us and the tech community and will help us make sure we're delivering excellent quality content.

Download a free PDF copy of this book

Thanks for purchasing this book!

Do you like to read on the go but are unable to carry your print books everywhere?

Is your eBook purchase not compatible with the device of your choice?

Don't worry, now with every Packt book you get a DRM-free PDF version of that book at no cost.

Read anywhere, any place, on any device. Search, copy, and paste code from your favorite technical books directly into your application.

The perks don't stop there, you can get exclusive access to discounts, newsletters, and great free content in your inbox daily

Follow these simple steps to get the benefits:

1. Scan the QR code or visit the link below

https://packt.link/free-ebook/9781804616659

2. Submit your proof of purchase

3. That's it! We'll send your free PDF and other benefits to your email directly

Part 1:
Legacy Estate Options

In order to understand the options for moving a legacy application to Azure, we must first look at the options for the various types of legacy estates.

This part contains the following chapters:

- *Chapter 1, Understanding Your Legacy Environment – the Modernization Journey*
- *Chapter 2, Strategies for Modernizing IBM and Unisys Mainframes*
- *Chapter 3, Midrange to Azure*
- *Chapter 4, Modernizing Legacy UNIX Systems*

1

Understanding Your Legacy Environment – the Modernization Journey

Before you choose the right option to move your legacy estate to a hyperscale cloud provider such as Azure, you need to understand the current environment of your legacy estate to know the options that might exist. This chapter will focus on the areas you should consider. Also, while many of the concepts that we will discuss can be applied to other hyperscale clouds, this book will focus on Azure. So, let's start with a list of topics that we will cover in this chapter to help you understand your current legacy estate:

- Current legacy hardware and operating system
- The current state of legacy applications
- What are the goals of moving to a hyperscale cloud such as Microsoft Azure?
- The need to choose a target architecture
- Consider your constraints
- How do you declare success for a legacy modernization to Azure?

Having been involved in many modernizations of legacy environments to Azure, understanding these fundamentals will both increase the likelihood of a successful modernization and provide you with an Azure estate that exceeds the capabilities of your current legacy environment.

Current legacy hardware and operating system

Many people think of mainframes when the term legacy estate is used. However, the term mainframe does not have the same meaning for all people. Additionally, there are non-mainframe systems that can either make up or surround a legacy estate. This section will focus on four types of legacy estates:

- IBM and Unisys mainframes
- IBM midrange
- Enterprise Unix
- Other legacy estates

IBM and Unisys mainframes

IBM mainframes have been around for over 50 years and have the largest share of the mainframe market. The current version is called the z series. Several operating systems can run on z hardware, including z/OS, z/VSE, z/TPF, and z/VM. Currently, IBM mainframes can also run Linux and specialty engines such as the **z Integrated Information Processor** (**zIIP**) and **Integrated Facility for Linux** (**IFL**). The most common operating environment we see is **z/OS**, which might also include zIIPs and IFLs in the same environment. This will be the type of IBM mainframe environment we will look at in this book for modernizing to Azure.

If you do have an IBM mainframe estate running z/OS, there are several other factors that you need to take into account before proceeding with an Azure cloud modernization. These include the mainframe size, features in use, and **Service Level Agreements** (**SLAs**). All of these factors need to be considered when modernizing to Azure.

The second largest mainframe market share belongs to Unisys. As with IBM mainframes, Unisys mainframes come in multiple versions. These include Libra (MCP) and Dorado (2200). To understand the differences, it is important to look at their history. Unisys is the result of a merger of two companies, Burroughs and Sperry. Libra comes from the Burroughs lineage. Dorado comes from the Sperry lineage. Keep in mind these are separate operating systems and require different modernization strategies.

As with IBM mainframes, you also need to take into account sizing, features, and SLAs.

IBM Midrange

For this book, we will consider IBM midrange systems to be POWER-based hardware that runs the iSeries OS. The IBM iSeries platform has been in existence for over 30 years and has included systems such as AS/400, System 36, and System 38. Many users of the iSeries platform also run mainframe systems. However, the iSeries is a distinct operating environment versus the IBM z series. As the name implies, **midrange systems** are typically smaller than mainframes. These systems may run mission-critical applications that require high performance. The iSeries operating system is tightly integrated with the POWER hardware to provide this high level of performance. Unlike the z systems, the iSeries is the only operating system for midrange. There are multiple versions of the iSeries in use today. But these share a common ancestry.

The POWER hardware can also run a Unix variant (AIX). This will be discussed separately in *Chapter 3*. Sizing, features, and SLAs are also factors for modernizing the iSeries.

Enterprise Unix

While there are several versions of Unix, for this book, we will concentrate on AIX, Solaris, and HP-UX. Unix systems are very different than mainframe and midrange environments. Unix systems were one of the first operating system environments that promised the option of Open Systems. However, as it turns out, the Unix of Open Systems meant the operating system could run on multiple hardware platforms, not that applications developed for one platform could run without modification on another platform. Unix systems are still widely used in several systems. However, each major variant requires a different approach.

IBM AIX systems on POWER hardware are widely used and provide high performance and availability. Solaris is also widely used. Solaris can currently run on x86 but also continues to run on Sparc hardware. HP-UX is still available on HP Epic (Itanium)-based hardware. Enterprise Unix systems were developed over 30 years ago but are more closely related to today's distributed and cloud systems. Enterprise Unix still has some legacy problems such as proprietary hardware/OS and agility.

Other legacy estates

Just to cover other legacy environments, there are several legacy operating systems and hardware. These include other z operating systems, such as z/TPF and z/VSE, other midrange systems such as Dec VAX, other mainframes such as Bull and NCR, and specialized operating systems such as HPE NonStop. While we will not go into detail about moving these environments to Azure, here are some options you can use for these environments that are similar to the areas we will cover:

- **z/TPF**: Similar to approaches for Enterprise Unix, but specialized throughput and availability
- **z/VSE**: Similar to approaches for z/OS
- **VAX**: Similar approach to Enterprise Unix
- **Bull and NCR**: Similar to the approach for Unisys
- **HPE NonStop**: Similar to z/OS, but with special attention to redundancy

These operating systems provide scale-up functionality similar to IBM and Unisys mainframes, and applications on these systems are typically written in legacy languages such as COBOL. This chapter will reference these systems in sections where features provided by them are discussed.

The current state of legacy applications

As with legacy hardware and operating systems, the current condition of applications that use your legacy estate is important to consider when you move to Azure. Here are the topics we will cover related to the current application estate:

- Scope of the legacy application estate
- Languages used in the current estate
- Third-party (COTS) applications
- Utilities (tools) used
- Operating system services
- Application-specific SLAs

Scope of the legacy application estate

While we recommend that you also do a detailed inventory and tools-based assessment of your legacy source code, there are some general rules of thumb you can use to scope out how to both accelerate modernization to Azure and take advantage of Azure services in the process.

For example, the total number of lines of source code is a good indicator of the effort needed to modernize to Azure. Lines of code are not necessarily a measure of complexity, as source code can often be converted using automated tools. However, lines of code will likely indicate the effort needed to test code once converted to Azure.

Another measure of complexity or effort deals with code interdependency. This is important as modules run independently and can usually be converted separately, thus simplifying the testing process. Otherwise, the interdependent code module will likely need to convert at the same time. For performance and testing, we advise using Azure tools such as Azure Monitor and Application Insights to assist with performance testing.

Another thing to consider is the possibility of unused code in your legacy estate. Many of your legacy applications were written across multiple decades. This means that your application may be full of *dead code* that is no longer used. Rather than converting and testing the dead code, running code analysis tools to identify code that is never called can accelerate modernization.

Identifying the data associated with an application you want to modernize to Azure is very important. For example, if you modernize the application code to Azure without also moving the associated data for that application, you are likely introducing latency that may not be acceptable to access that data.

Finally, identifying if you have all the source code and the current version is extremely important. A lack of source code will limit the modernization options you have.

Languages used in the current estate

Since legacy systems were developed over decades, there may be some obscure development languages that it might be hard to find cloud versions for or are simply no longer supported. More commonly, however, you will find older languages such as Cobol or PL/I that are not supported with common tools and IDEs in Azure without third-party **Independent Software Vendor** (**ISV**) compiler software.

In particular, if your goal is to move to Azure-native managed code solutions for your legacy applications, you may find the need to integrate your managed code (for example, Java or C#) with unmanaged code in C/C++ or other managed executables.

Third-party (COTS) applications

Another possibility is the presence of third-party software in the legacy environment. Since most legacy applications not written in Java or C# (or another managed language) are not compatible with Azure, we need to find whether the third-party ISV provides a version that is compatible with Azure. For example, a **commercial off-the-shelf** (**COTS**) application that currently runs on AIX may have a version that runs on x86 Linux. In that case, moving the COTS application to Azure is typically not a technical problem. There may be licensing issues that need to be addressed. But it's not a technical problem. If that is not the case, you will likely need to find an alternative for the COTS application. In some cases, such as POWER platforms, running the COTS application as-is on Azure might be an option. Otherwise, an alternative application that runs in Azure will need to be found.

Utilities (tools) used by applications

Many applications depend on utilities that provide services by the operating system to function. This includes things such as queuing, transactions, data replication, scheduling, monitoring, backup, and printing. For example, many applications use queuing provided by MQSeries. When moving to Azure, you will need to decide if you wish to continue to use MQ, which is available in Azure, or use a native Azure queuing server such as Service Bus.

We recommend mapping all the utilities currently used by your legacy application with utilities available in Azure. In some cases, these may be native Azure services such as Service Bus, Azure Monitor, and Azure Data Factory. In other cases, third-party tools might be required for things such as printing and virtual tape.

Operating system services

A special type of utility deals with operating system services. This is particularly true for IBM mainframes and midrange. This includes features on the mainframe such as CICS, IMS DC, and Coupling Facility. When we look at these features further, CICS and IMS DC are really like application servers on the mainframe. So, on Azure, we need to look at things such as App Services or Tomcat hosted in Azure Kubernetes Service to provide similar functionality.

On the midrange, the operating system provides tight coupling with the database for applications, and integration with scripting using CL. For this type of functionality, we need to extract these functions from the applications and put them in the database, scripting language, or other utility functions on Azure.

Application-specific SLAs

The SLAs provided natively by Azure for **virtual machines** (**VMs**) and data solutions will likely meet the availability and recovery requirements for most legacy systems. However, there are some specific cases where that may not be the case.

Applications that were developed by legacy programmers often depend on proprietary features of the operating system to achieve high availability and resilience. An example would be COBOL applications run in a CICSPlex and use Db2 that runs in a Parallel Sysplex. In this type of scenario, the applications themselves are unaware of the underlying system software that provides application resilience and limits data loss. Application redundancy and data resilience in cloud systems such as Azure typically rely on scale-out redundancy to meet similar availability. The scale-out approach usually relies on the application to be able to restart. That logic might not be present in the legacy code being converted.

There are mitigating approaches that can be followed, such as using a caching service such as Redis Cache on Azure. However, either the ISV software being used for the application layer, or the application code itself, would need to be Redis Cache-aware.

Mainframe applications were typically developed as monoliths (single system, not distributed). This section covered the various requirements Azure needs to provide to offer the same type of robustness and features.

What are the goals of moving to a hyperscale cloud such as Microsoft Azure?

So, what are the reasons behind deciding on a migration or rewrite of an application or ecosystem of applications to Microsoft Azure? The answer to that question will no doubt be different for everyone. Typically, the reasons can be broken down into three general categories.

Cost-related

The following are the cost-related factors:

- **Elastic scalability**: Legacy systems were designed to run 24/7 and to support peak volumes, you were forced to design the supporting architecture appropriately. This meant that most architectures were over-provisioned. In Microsoft Azure, you can right-size your architecture and grow it as needed, either vertically or horizontally, due to the elastic nature of the cloud. This elasticity applies to the main elements of any data center – that is, its compute, storage, and networking. You pay for what you need and use and (more importantly) you do not pay for what you do not need. This can allow significant cost savings.

- **CapEx versus OpEx**: Traditionally, for legacy applications, a large **capital expense (CapEx)** was necessary to provision a data center. This meant a large outlay of cash for assets that started to immediately depreciate. In a cloud model such as Microsoft Azure, these costs are transformed into **operating expense (OpEx)**. Operating costs are the day-to-day expenses necessary to run a business. Essentially, hyperscalers are similar to utilities, which offer a significant advantage since you do not need to depreciate assets and deal with obsolete hardware.

Application development-related

The following are the application development-related factors:

- **Modern developers**: COBOL is a procedural language that was created more than 50 years ago. Many mainframe applications were developed in COBOL and also Assembler. Since these languages are so old, there is no provision for things that are inherent in modern languages such as polymorphism and object orientation. Also, aside from a few colleges that teach COBOL, the population of COBOL developers in the world is decreasing. More and more developers are being trained in modern languages such as Java and C#. Moving to Azure allows you to leverage these new developers so that you can maintain these applications.

- **Modern DevOps**: The process of creating applications has come a long way since the use of tools such as **Time Sharing Option (TSO)/Interactive System Productivity Facility (ISPF)**, which was the solution supplied by IBM for mainframe application development. In today's cloud-based world, application pipeline management tools and repositories such as Git are the norm. Solutions such as Azure DevOps allow for the streamlined development of modern applications by provisioning developer services that allow teams to work and collaborate on code development and deploy applications.

- **Modern architectures**: We will explore this in more depth in *Chapter 6* and *Chapter 7*, but many modernization efforts today are being driven simply by the need for your business to be more agile and leverage modern cloud architectures such as microservices. This is especially important in competitive industries such as finance, transportation, and retail. Modern architectures allow businesses to respond more quickly to changing business environments. In government settings, they allow government entities to respond more quickly to situations such as COVID-19 and new regulations.

Hardware-related

Obsolete hardware is the major hardware-related factor. In my travels focusing on legacy modernization, I have been surprised by how many companies want to move to Microsoft Azure simply to get off aging hardware. There have been numerous occasions when customers have told me that they had to resort to eBay to buy hardware that was not manufactured and supported anymore. The only option was to modernize or be at the mercy of aging hardware. When you think about it, this was probably to be expected. These were very reliable systems. They just ran the business. Management changed and the business environment changed, but the legacy systems kept running until someone asked if they should be modernized.

So, cost, application development, and hardware are the key drivers for modernizing a legacy application. As you consider where you might be in the drivers listed earlier, keep in mind that there are many paths to modernization. This is the goal and objective of this book – to show you which path might be the best for the current situation you are in.

The need to choose a target architecture

Just as important as understanding the platform that you are moving from, you have to have a clear vision of what your target platform needs to look like. In Azure, this target platform will be quite different from the legacy platform, mainly because it will all be virtual. The need to define a target architecture for your legacy workloads is the first step in laying the foundation of your virtual data center in the cloud. The compute, networking, storage, and other ancillary services need to be thought out and defined prescriptively in Azure if you wish to have a scalable and manageable environment. An example here might explain this concept.

In Azure, one of the fundamental containers for resources is an **Azure subscription**. One approach that some have used when defining their Azure environment is to create one subscription and then create all the necessary artifacts and assets within that subscription. There are some inherent problems with this approach when it comes to manageability and governance. On the opposite end of the spectrum, there is the approach of overusing subscriptions since they provide such a nice security and billing boundary. The correct approach lies somewhere in the middle. This is where **Azure Landing Zones** can be very useful.

Azure Landing Zones

We will discuss Azure Landing Zones later in this book, in *Chapter 6*. For now, it is good to know that an **Azure Landing Zone** (**ALZ**) allows the Azure architect to create a prescriptive architecture typically defined in an **Azure Resource Manager** (**ARM**) template. This approach allows for a scalable and modular implementation that is highly agile and can be deployed for different purposes. A good example is standing up a replica of production for QA testing or possible training.

There are essentially two types of landing zones:

- **Platform Landing Zones**, which are typically used to provide centralized services such as networking and identity. These services will be used by various workloads and applications.

- **Application Landing Zones**, which are typically application-specific and are used, for example, as a way to ensure that group policies are being applied correctly for a particular application group.

For this book, we will focus on **Application Landing Zones** as our intent will be to migrate legacy applications.

One of the first questions that needs to be answered when you are defining the target environment is, *What are my target application containers?* Should I use virtual machines, typically called **Infrastructure as a Service** (**IaaS**), Kubernetes, or the more cloud-native microservices? There are pros and cons to each.

Infrastructure as a Service (IaaS)

IaaS uses Azure Virtual Machines as the application container. VMs have been leveraged for years both on-premises and in the cloud. Virtual machines are a virtual definition and implementation of a physical machine and as such, they require an OS and application software. This means that they take more time to deploy and manage than, say, something such as Kubernetes. Having said that, virtual machines can be grouped in application clusters within Azure and can horizontally scale. We will dive into this later in *Chapter 6*.

Here is an example of an architecture for multiple Mainframe workloads migrated to Azure. You can see that the primary site is replicated to a paired region where **Azure Site Recovery** (**ASR**) keeps the application tier virtual machines in sync, while Microsoft SQL Server Always On provides fault tolerance for the data tier:

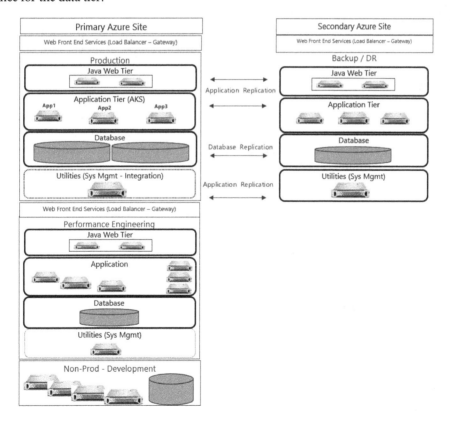

Figure 1.1 – High-level Azure architecture for legacy systems

Azure Kubernetes Service (AKS) and microservices

Microservices decompose the applications running on virtual machines into independent modules that are loosely coupled and distributed. The benefits are applications that are much more scalable and agile but also more complex to manage unless you have a managed service such as Kubernetes. In Azure, this service is called **Azure Kubernetes Service (AKS)** and provides a way to orchestrate and define relationships between components.

Azure Functions and Service Fabric

There are other deployment vehicles for microservices in Azure, such as Azure Functions and Service Fabric. They offer similar benefits with some variations. We will discuss these later in *Chapter 6*.

Now that we have looked at the need to define a target architecture, let's understand the constraints.

Consider your constraints

Most legacy modernization projects for moving to Azure have constraints, meaning there are non-technical constraints, such as time and money, or technical, meaning things such as scalability and performance that have to be addressed. This section will cover these constraints.

Time constraints

Many legacy migrations are bounded by time constraints. With mainframes, this is the process of renewing the mainframe hardware, which typically happens every 3 years. This can help define the time frame necessary for the migration to be completed. It is very important to understand the size of the workloads and the number of lines of code that need to be migrated to meet this time frame. There are various types of time constraints, depending on the industry you are in:

- **Hardware renewal constraints**: As mentioned earlier, these are for support and maintenance and are typically 3 to 5 years.
- **Hardware supportability constraints**: For most hardware vendors, they only support previous versions of the hardware back to a certain version. They vary by vendor.

Resource constraints

Even though there are large parts of a migration that can be automated, another bounding factor is the number of resources necessary for the migration. This is extremely important when it comes to testing, both smoke testing and user acceptance testing. Human resources such as developers and testers are critical, as are resources such as testing and development environments.

Funding constraints

And we cannot forget about funding the migration. There are multiple costs to consider that need to be aligned with the funding to make sure the migration is even worth the effort. Perhaps buying and implementing a COTS or **Software as a Service (SaaS)** solution would be more efficient:

- **The cost of the Azure infrastructure**: This cost is relatively easy to calculate. It involves mapping the compute from the legacy system to Azure vCPUs, as well as the storage and networking to the appropriate Azure artifact.

- **The third-party software costs**: Depending on the migration software vendor, there will be additional software costs for additional components of the solution. A good example is if the legacy solution leverages a scheduler on the legacy platform; a corresponding scheduler will need to be implemented in Azure.

- **The migration cost**: Even though automation can be used to minimize the number of people necessary to migrate a solution, the actual migration, be it a rehost, refactor, or rewrite, will be a significant part of the cost of the migration.

Now that we've looked at constraints, we will look at defining success for a modernization project.

How do you declare success for a legacy modernization to Azure?

In this chapter, we reviewed several different states that might comprise your current legacy environment. We also reviewed the goals you might have for an Azure solution and the constraints you currently face. Now, it's time to discuss setting up realistic goals, or possibly multiple steps that might be needed to achieve those goals. The idea here is to have early success so that you can build on it. Legacy modernization projects can seem quite daunting, even when there are few or no technical hurdles that need to be overcome. That is why we believe early success makes the daunting task now look achievable.

To help with this, we believe there are several things you should consider.

Identify the first workload to modernize to Azure

This may sound obvious, but there may be several factors you might want to consider, such as the following:

- Will the first application provide a template for replicating similar applications that use Azure features, such as AKS or ASR? In other words, pick an application that has features in common with other applications you want to modernize. Do not choose a one-off.

- Find an application that can be done in a reasonable time frame and achieve a reasonable ROI. These are great measures of success.

- Be careful not to start with the most difficult application move. This is a recipe for failure.

- Make sure business continuity is a key part of the modernization to Azure process.

- Understand the effect latency might have on modernization. Unacceptable latency is another recipe for failure.

- Make sure that the team responsible for the modernization to Azure has the right training to take the application you choose to Azure.

- Make sure you have the proper testing and validation tools and reports to verify success. This could include the use of Azure tools such as Azure Monitor and Application Insights but may also include third-party analysis and benchmark tools.

Determine if modernization can be a multi-step process

Keep in mind that modernization can be a multiple-step process. For example, if you first need to move from on-premises to Azure, moving an application without changing the language might be a good first step. The second step might be to refactor the legacy language to a modern language. The third step might be creating an Azure-native solution.

While it is possible to move directly to an Azure-native solution, doing so will likely require both architecture and source code changes that will take extra time and require more testing. Also, going straight to an Azure-native implementation may be most successful with smaller application modernizations. This makes the process more manageable.

Establish hybrid and integration requirements

Most legacy applications are not standalone and self-contained. There is usually a need to both integrate and interoperate with other legacy applications or applications already on Azure. This means that a hybrid strategy for integration and interoperability is a key feature for success. In looking at the right application to modernize to Azure, there should be the right balance of hybrid or integration needs, along with the ability to segregate functionally so that things such as latency do not become a big issue. Fortunately, Azure provides tools that make creating a hybrid environment practical. They include the following:

- **Azure Data Factory**: For data ingestion and transformation

- **Logic Apps**: For low-code data integration

- **Power Apps**: To extend hybrid functionality with low-code solutions

- **Azure Service Bus**: For integration with MQSeries or other message environments

- **Azure Event Hubs and Event Stream**: To enable hybrid event logic

Establish repeatable processes

As a final thought for this chapter, an initial modernization success should also be a recipe for a repeatable process. While not all applications can be modernized to Azure in the same way, there are probably applications that can be categorized as being similar in their approach to Azure. Success in an initial modernization should provide a temple for other applications to follow.

From our experiences, repeatable processes can be achieved in modernization to Azure by first categorizing the application candidates into approaches that are appreciated for each category. Here are some examples:

- If an application currently meets business needs and requires minimal maintenance, then perhaps a rehost of that application or the use of an emulator would be the best option while keeping the same legacy language. This could provide a template for rehosting to Azure for your legacy estate.

- If an application currently meets business requirements but requires a lot of maintenance to the source code, it might be better to refactor (convert it into another language) to allow non-legacy programmers to maintain the application. This does not mean a complete re-architecture of the application, but it will allow you to maintain the application in a modern language. Any example would be refactoring to either Java or C#.

- Finally, if the application needs an overhaul for both business functionality and deployment. This type of application would be best reimagined as an Azure-native architecture.

A single approach may not be appreciated for all applications you want to modernize to Azure. In this case, you may want to create parallel tracks to achieve the optimal paths to Azure.

Now that we've looked at how to leverage a successful modernization project to lead to future successful modernizations, we will summarize this chapter.

Summary

In this chapter, at a high level, we looked at the overall process of defining, planning, executing, and reusing modernization projects for moving legacy applications to Azure. For the remainder of this book, we will cover additional details regarding each section of this chapter, starting with IBM and Unisys mainframes.

2

Strategies for Modernizing IBM and Unisys Mainframes

As we mentioned in *Chapter 1*, when we refer to mainframes in this book, that covers the IBM Z series and the Unisys Dorado and Libra series. This chapter will cover at an in-depth level the strategies to modernize these mainframes to Azure. Here is a list of topics we will cover:

- IBM mainframes:

  - **z/OS**—The most common IBM mainframe operating system

  - **z/Virtual Storage Extended (z/VSE)**—Similar to z/OS, but usually for smaller workloads

  - **z/VM**—Z-series virtualization

  - **z/Transaction Processing Facility (z/TPF)**—For high-volume transaction workloads

  - **z/Linux**—Linux that can run under z/VM

  - Specialty engines—**Integrated Facility for Linux (IFL)** and **z Integrated Information Processor (zIIP)**

- Unisys:

  - **Libra**—From the Burroughs **Master Control Program (MCP)** line

  - **Dorado**—From the Sperry OS 2200 operating system

IBM mainframes

To start with, we think it's important to understand a little about the underlying hardware. The physical hardware instance, which is sometimes called a **Central Electronics Complex (CEC)**, can have one or more compute engines, such as the **general-purpose (GP)** zIIP or IFL, and each compute engine can run one or more operating systems. Additionally, each CEC can be coupled to another CEC to provide a tightly coupled scale-out environment, as illustrated here:

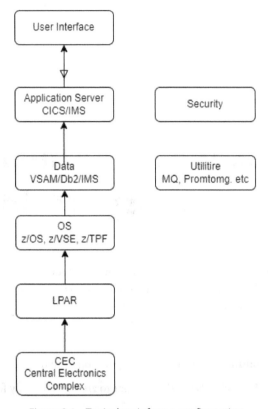

Figure 2.1 – Typical mainframe configuration

This is what a typical mainframe configuration looks like.

z/OS – The most common IBM mainframe operating system

When most people think of mainframes, they are thinking of z/OS running on a GP processor on Z-series hardware. This is the operating system commonly used by banks, insurance companies, large retailers, and government agencies. The first step toward modernizing a z/OS system to Azure is to understand a little about z/OS.

To start with, you can think of z/OS as an operating system for a **monolithic server** that combines many of the features that Azure offers either as native Azure services or with **Infrastructure as a Service (IaaS)** VMs. These include the following:

- **UI**: Classic green screen or web
- **Application server: Customer Information Control System (CICS)** or **Information Management System Transaction Manager (IMS TM)**
- **Batch processing: Job Control Language (JCL)** and scheduling

- **Database**: Db2, IMS DB, **Virtual Storage Access Method (VSAM)**, and so on

- **Utilities**: Queuing, backup, printing, and so on

- **Security**: **Resource Access Control Facility (RACF)**, **Access Control Facility 2 (ACF2)**, Top Secret, encryption, and so on

Additionally, there can be multiple instances of z/OS that are running in **logical partitions (LPARs)** on a given CEC. LPARs play a similar role to separate VMs that run in Azure. For example, there might be one LPAR for production and a separate LPAR for dev/test. In Azure, you might have separate VMs for the same purpose, as illustrated here:

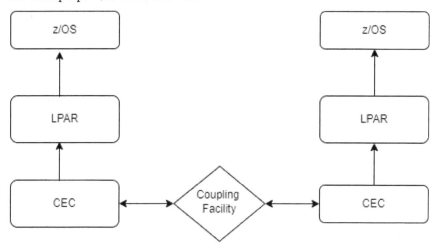

Figure 2.2 – LPARs, CECs, and coupling facility

Let's look at the six aforementioned areas and how each area might work in Azure.

UI

z/OS mainframes can have character-based UIs (classic green screen) or a web browser UI using WebSphere for z/OS. Additionally, since the mainframe can host Linux either in a VM or IFL, the UI for the application may not be running on z/OS.

Character-based UIs might seem a bit archaic, but they are still commonly used and even preferred by experienced users. Character interfaces are extremely fast. So, even though the character UI might be harder to learn and master, the character UI provides fast response times. This is particularly true in applications with customer service agents who have a lot of interaction with the customer, such as with reservation systems. With Azure, TN270 emulation can provide this type of green-screen interface. Microsoft has a TN3270 emulation feature with our Host Integration Server product that can run in Azure. Additionally, there are many third-party TN3270 emulators. An overview is provided in the following diagram:

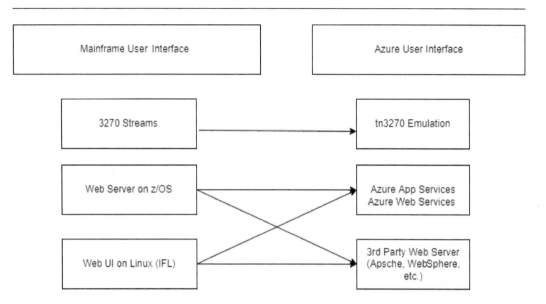

Figure 2.3 – Mainframe UI versus Azure UI

If the z/OS system already uses a web interface, then this could be modernized to Azure using any of the following services:

- Azure App Service

- Azure Web Apps

- WebSphere for Linux running on a VM in Azure

- Apache or other open source web-based UI running in a container on Azure

Your decision will be determined by choosing the right method for UI modernization in Azure that not only provides the UI features the application needs but also fits the overall method that other applications in your environment use for UIs in Azure.

Application server

Some mainframe applications run entirely in batch (*no UI*) mode. For these, an application server is probably not needed. However, if your z/OS environment has an online and associated UI, we need to discuss how to move the online UI to Azure. Most z/OS applications with an online UI use either CICS or IMS TM for the online UI. You can think of both of these systems as the z/OS mainframe's application server. While CICS and IMS TM both perform similar functions, they approach their function as application servers in a very different manner. Each needs a strategy to migrate to Azure.

CICS

You can think of CICS as an application server that hosts programs that interact with CICS using specific verbs and syntax that are integrated into the programming language used to develop the application. The most common z/OS programming languages that use CICS include **Common Business Oriented Language** (**COBOL**), **Programming Language/I** (**PL/I**), and Assembler. CICS manages the state for a UI session through a feature called the COMMAREA. This allows CICS to pass information from one application program to another without requiring a separate session (*thread*). This is not dissimilar to how web servers manage the state of browser applications. CICS can also coordinate transactions between multiple data providers.

Modernizing CICS to Azure will require using a third-party service that supports CICS syntax or modifying the application source to remove the CICS-specific syntax and use syntax native to Azure services or third-party application servers.

A typical CICS configuration is presented in the following diagram:

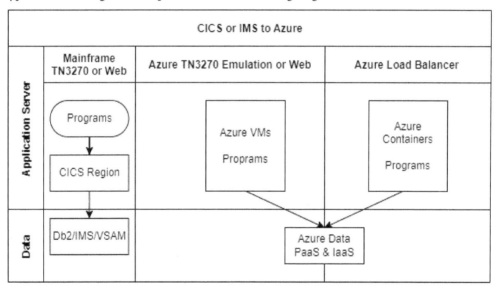

Figure 2.4 – Typical CICS configuration

The choice of which path you want to take will depend on a number of factors, such as the following:

- Do you want to retain the legacy source code, such as COBOL and PL/I?
- Do you want to refactor or rewrite the application source code?
- Do you have a standard for application servers within the business?

Fortunately, Azure provides options for each of these.

IMS TM

IMS TM provides a message-based application server that programs can use with IMS verbs and syntax. As with CICS, with IMS TM you will need to decide whether you want to keep IMS syntax or modify the code using either Azure native services or third-party application services. The difference with CICS is the message-and-queue nature of IMS.

There are third-party **independent software vendor** (**ISV**) solutions for Azure that support IMS syntax. However, if you want to remove the IMS syntax, you will need to use some type of message/eventing system as an alternative. This might include the following:

- Azure Service Bus for messaging and queuing

- An eventing system such as Azure Event Hubs or Kafka

- An IaaS-based queuing system, such as MQSeries or IaaS SQL Server with Service Broker

As with CICS, the decision on which path to take primarily deals with the type of system you want to have on Azure.

The big difference is the dependence on messaging and queuing for the applications that currently use IMS TM.

Batch processing – JCL and scheduling

IBM z/OS mainframes will typically have a daily batch set of programs to run in addition to the online **user interface** (**UI**). There can be many different applications with several steps each that need to run in a certain order. To manage this, z/OS systems use a scheduling system to initiate jobs that run applications, and JCL to control steps within a job's execution.

For job scheduling, you will need to decide whether you want to keep a scheduling system that is similar to what is used on the mainframe or change your approach to be more Azure-native. If you want to keep a mainframe-style scheduling system, you will need to find a third-party ISV that provides this type of solution that runs on distributed systems. If you wish to go with Azure native, you can look at the Azure Scheduler service or an event-based solution such as Azure Event Hubs or Kafka. However, keep in mind the move to an Azure-native scheduler will require design work to fit the new paradigm.

For JCL, the decision you need to make is whether you would like to retain the z/OS JCL or rewrite the JCL to Azure-compatible scripts. With that said, z/OS systems could have several thousands of lines from JCL. If you want to keep JCL, that will require a third-party ISV that has a JCL interpreter. These are usually the same ISVs that provide CICS and IMS solutions that can run on Azure. Moving to a native Azure scripting solution will require a transformation. Tools can automate this to some degree, but there will likely still be some manual effort.

Database

There are several options for data on the mainframe. For this chapter, we will look at the following:

- **Db2 for z/OS**: Relational database for z/OS

- **IMS DB**: Hierarchical database for z/OS

- **VSAM**: File-based data records that includes indexed, relative, and sequential access

- **Adabas/Datacom/Integrated Data Management System (IDMS)**: Third-party solutions

The options for moving z/OS data to Azure include the following:

- **Db2**:

  - Azure PaaS (**Platform as a Service**) data services such as SQL DB, PostgreSQL, and **SQL Managed Instance (SQL MI)**

  - IaaS databases, such as SQL Server, **Db2 for Linux, Unix, and Windows (Db2 LUW)**, Oracle, and PostgreSQL

- **IMS**: Cosmos DB and SQL Server

- **VSAM**: SQL Server, Cosmos DB, and other relational databases

- **Other databases**: Adabas on Linux in Azure, SQL PaaS, and other IaaS relational databases

Our recommendation is to look for Azure-native data solutions such as SQL DB, SQL MI, and Cosmos DB whenever possible. An overview of the options is provided in the following diagram:

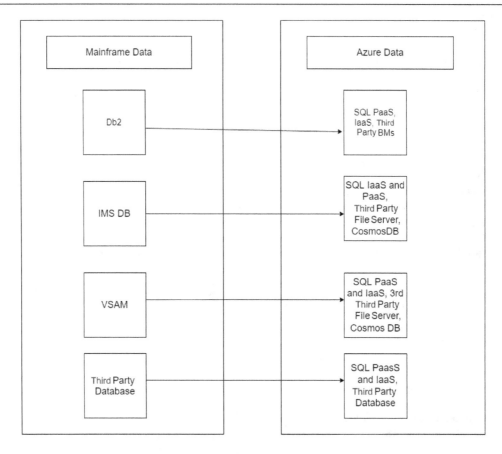

Figure 2.5 – Mainframe data comparisons

The primary reason for selecting Azure-native data solutions is the management that PaaS services provide. From our point of view, an IaaS database requires more personnel than a PaaS database.

Utilities

IBM mainframes running z/OS usually include a number of utilities that we need to take into account when we modernize to Azure. These include the following utility services:

- **MQSeries**: Use MQSeries on Azure or Azure Service Bus
- **Virtual tape**: Third-party vendors with virtual tapes to Azure Blob storage
- **Scheduling**: Re-architect to Azure native scheduler or use third-party scheduler
- **Printing**: Third-party software or .pdf conversion
- **Backup**: Azure-native or third-party Azure solution

Security

IBM z/OS mainframes provide security in several ways, including user authentication, user authorization, certificates, and encryption. The following table lists a mapping of these features for z/OS versus Azure:

Security Feature	z/OS	Azure	Comments
Authentication	RACF ACF2 Top Secret	**Azure Active Directory (AAD)** **Active Directory (AD)**	In Azure, customers might use a combination of AAD and AD
Authorization	Same as for authentication, along with integration with CICS and IMS TM	AAD/AD with transaction-level security provided by third-party software	
Certificates	Hardware- and user-based	Hardware- and user-based	
Encryption	Hardware- and software-based	Hardware- and software-based	Hardware-based requires a crypto processor on mainframe or **hardware security module (HSM)** VM support on Azure

Table 2.1 – Mapping security capabilities

While both z/OS and Azure can provide similar functionality for security, this might not be implemented the same way. For example, CICS on the mainframe integrates at the transaction level with RACF. However, when using a third-party ISV CICS solution for Azure, you may need to use the ISV's transaction-level security rather than relying entirely on AAD/AD for authorization.

z/VSE – Similar to z/OS, but usually for smaller workloads

As the subsection's title implies, z/VSE is typically used for smaller workloads than z/OS. As with z/OS, z/VSE (or VSE) has been around for 50+ years. z/VSE also runs both online and batch workloads using languages such as COBOL, PL/I, and Assembler. That is where the similarity ends. For example, the JCL for z/VSE is not compatible with the JCL for z/OS. Also, since z/VSE is for a smaller market segment, most ISV solutions that assist in modernizing z/OS applications will not work for z/VSE applications.

For the reasons we just discussed, we recommend either refactoring the application source code or re-architecting the entire application when moving from z/VSE to Azure. With that in mind, you will need to look at how the application is composed and choose the best path for hosting that application, just as was recommended for z/OS. As a reminder, here is a list of areas you should consider:

- **UI**: Classic green screen or web
- Application server
- Job scheduling and job control
- Database
- **Utilities**: Queuing, backup, printing, and so on
- Security

Many z/VSE systems are primarily batch based. This makes converting to a scheduling system on Azure and converting the JCL of particular importance.

z/VM – Z-series virtualization

z/VM is virtualization for a mainframe beyond what you can do with LPARs to separate operating environments. For example, you can use z/VM to run separate instances of z/OS and even z/VSE, z/TPF, or even z/Linux.

The function that z/VM provides can be run by either VMs or multiple instances of services on Azure. In other words, if you need two separate operating system instances and two databases, you can create two VMs and two instances of Azure SQL DB.

z/TPF – For high-volume transaction workloads

z/TPF is a highly optimized operating system for high-volume transactions, such as credit card payments and travel reservations. z/TPF can also take advantage of Z-series coupling facility features for tightly coupled scale-out scaling. Additionally, z/TPF only supports a subset of programming languages and data options, as illustrated here:

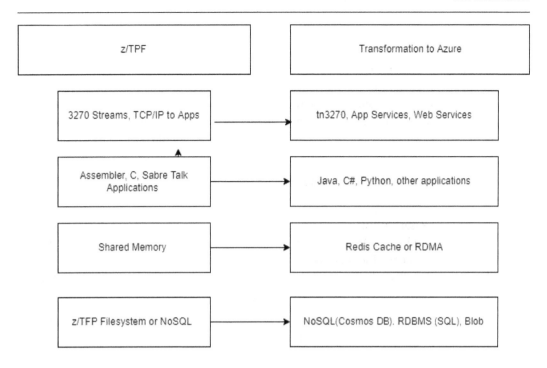

Figure 2.6 – Mapping z/TPF to Azure

All of this means that moving z/TPF applications to Azure will require applications to Azure services that provide equivalent performance and **service-level agreements (SLAs)**. Our recommendation is to understand the scope of the effort for moving this type of system to Azure and engage appropriate personnel for the next step.

z/Linux – Linux that can run under z/VM

If you have Linux running either on z/VM or IFL, that can typically be easily migrated to Azure on a VM. Some issues may require addressing, such as big endian (on mainframe) to little endian (on Azure), and any code page translation issues. That said, moving Linux on the mainframe to Linux on Azure is usually a fairly straightforward process.

Specialty engines – IFL and zIIP

IFL and zIIP are specialty engines that IBM mainframes offer. The IFL engine runs Linux, which we covered in the previous subsection, while the zIIP engine runs Java services. The data services are covered in the *Database* subsection of the *IBM mainframes* section of this chapter. The Java services are similar to the migration of the *Application server* subsection of this chapter for GP processors.

We've finished looking at IBM mainframes. We will now look at Unisys mainframes.

Unisys

Unisys was formed from the merger of two companies in 1986: Burroughs and Sperry. The name is derived from three words—*united*, *information*, and *systems*. While the original intention was to unify the two operating systems, they still exist today as separate operating systems. Each of the merged companies had a line of mainframes, and these separate types of systems still exist today after over 30 years. Libra is the new name for the Burroughs product line, and Dorado is the new name for the Sperry line.

As these are different types of mainframes than IBM, we will look at them separately for modernizing to Azure.

Unlike IBM with the Z-series systems, Unisys has taken a different approach to modernizing its mainframe customers. It has created a hardware emulator that runs on x86 platforms for both the Burroughs and Sperry systems. The collective product name for this offering is **ClearPath Forward**. There is ClearPath Forward for the Burroughs system, **Libra**, and ClearPath Forward for the Sperry system, known as **Dorado**.

This means that a Libra system can be hosted on a VM in Azure using ClearPath Forward, assuming the VM meets the specs provided by Unisys and the VM has its compute and networking systems configured properly.

This emulation approach makes the modernization journey much easier for customers, with a true lift-and-shift approach whereby the customer can take the partitions that run the workloads and copy them to ClearPath Forward, and they run without any recompilation. Let's take a look at each service.

Libra – From the Burroughs MCP line

The Libra MCP operating system originated over 40 years ago and is still used today by financial services and other customers. In ClearPath Forward, the entire operating system is emulated in a properly configured Windows VM.

As with the IBM Z series, the Unisys Libra operating systems can be viewed as servers that provide similar services to z/OS. These services and functions provide services for online user screens, batch processing, scheduling, and other necessary utilities, as outlined here:

- **UI**: Classic green screen or web
- Application server
- Job scheduling and job control
- Database
- **Utilities**: Queuing, backup, printing, and so on
- Security

Other options for modernizing are to either rehost the Libra system on VMs or containers in Azure using third-party software or refactor/rewrite the application to a language that can run on Windows. We will look at the pros and cons of the different approaches to modernizing later in this book, but for now, let's look at why you would want to take one approach over the other:

- **Lift and shift to ClearPath Forward Libra**: This is the option with the least risk. What you have for your mainframe now, you will have in the ClearPath environment in Azure. The only difference will be the cost and the fact that your system is now in a public cloud where you can more easily leverage things such as cloud-based analytics.

- **Rehosting**: This entails recompiling the COBOL programming language and other applications to x86. It has more risk because you have to deal with potential differences in the way the compiler might work, but the benefit would be that the performance will be better because you are not having to run in an emulated hardware environment. With rehosting, your languages would still be the same, but in a more modern **integrated development environment** (IDE) such as Visual Studio or Eclipse.

- **Refactoring**: This is very similar to rehosting except that you end up in a modern language such as Java or C#. It has more risk than rehosting typically, but if you are looking to have developers who are more familiar with modern languages such as Java perform the maintenance to the applications, this might be a better fit.

UI

As with the IBM z/OS systems, if you are *old school* and still want to manage the system via a command line, you can do that via your favorite terminal emulator. However, for ClearPath Forward for Libra systems, the majority of management and administration activities are accomplished through the **ClearPath ePortal** for MCP. ClearPath ePortal is an application that provides a holistic view and an end-to-end solution for monitoring, managing, and administrating ClearPath MCP applications. It allows you to integrate them with business partners and also provides a way to integrate with modern mobile applications This allows you to connect your migrated legacy applications to mobile applications and modern web services.

Application server

The application server functionality is built into the core offering of the solution. It is offered in three components: Application Enrichment Services, Product Implementation Services, and Managed Services. Management, scheduling, and job control for workloads within the ClearPath environment are provided with the Workload Management for ClearPath MCP component.

Database

Data-tier functionality is provided by Enterprise Database Server for ClearPath MCP. It provides a high-volume transaction-processing database environment for installation of any size. It can accommodate relational, flat, and hierarchical data models. It is supported by Unisys Data Exchange, which is a data integration solution that can perform **extract, transform, load** (ETL) activities to and from various data sources.

Utilities – Queuing, backup, printing, and so on

Utility services are provided within the ClearPath Forward environment itself through the **ClearPath Forward Fabric Manager** UI. There are various components for adding additional ClearPath Forward platforms (that, processing environments). Within a platform are partitions, which can also be added via the **Fabric Manager** UI. Additionally, there is also the ClearPath MCP IDE for Eclipse platform, which provides an IDE for building ClearPath Forward composite applications.

Security

Unisys ClearPath Forward for Libra Servers is a self-contained processing environment that has its own built-in security, which is very secure. It runs on the Microsoft Windows platform, so the security of the host environment is managed via what is provided in Windows.

You can see a legacy Burroughs system and what that same system would look like in ClearPath Forward here:

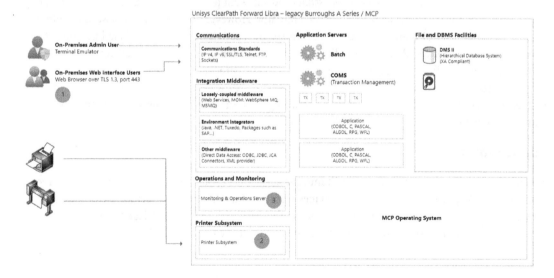

Figure 2.7 – Pre-migration

And let us look at it post-migration:

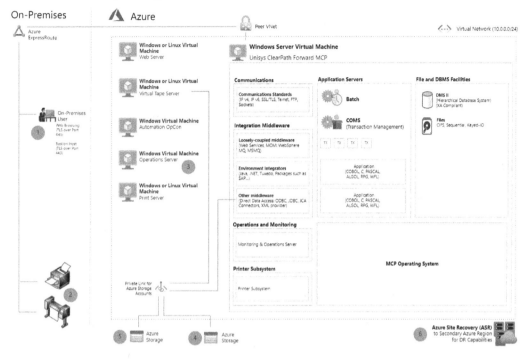

Figure 2.8 – Post-migration

Now, let us look at Dorado.

Dorado – From the Sperry OS/2200 operating system

Let us now delve into Dorado. As with the Libra offering, this is a solution for a true lift and shift for Unisys/Sperry mainframe systems. This means there is no recompilation necessary, which reduces the risk, and the transition to Azure takes place by moving the workloads from the Unisys mainframe to a specially configured Linux VM in Azure.

UI

For ClearPath Forward for Dorado systems, just as with the Libra offering, the majority of management and administration activities are accomplished through the ClearPath ePortal for OS/ 2200. ClearPath ePortal is a Windows-based application that provides similar functionality to the MCP version. It allows you to modernize ClearPath OS/2000 applications, extending them to reach new applications through mobile and web services. And, just as with MCP, this allows you to connect your migrated legacy applications to mobile applications and modern web services.

In addition to the ClearPath ePortal for OS/2000, there is an additional GUI web-based management and administration tool that allows the management of users, groups, and various system administration tasks. It also has built-in reporting as well.

Application server

The application server functionality is built into the core offering of the solution as well. It is offered in three components: Application Enrichment Services, Product Implementation Services, and Managed Services. Management, scheduling, and job control for workloads within the ClearPath environment are provided with the Workload Management for MCP component.

Database

The data tier is provided by the ClearPath implementation of the Unisys **Universal Data System** (**UDS**). This provides a common control structure for different data models such as relational, hierarchical, and flat. And similar to Enterprise Database Server for ClearPath MCP, it is supported by Unisys Data Exchange, which is a data integration solution that can perform ETL activities to and from various data sources.

Utilities – Queuing, backup, printing, and so on

The queuing middleware necessary for application-to-application communication is provided by Unisys ClearPath OS 2200 Processor. This requires a special partition within the system.

Security

As with the Libra offering, Unisys ClearPath Forward for Libra Servers is a self-contained processing environment that has its own built-in security, which is very secure. It runs on the Microsoft Windows platform, so the security of the host environment is managed via what is provided in Windows.

We've now looked at Unisys mainframes too. We will now summarize this chapter.

Summary

So, as you can see, when you hear someone say that they have a *mainframe*, it can mean different things. Sometimes, they may actually be referring to a mid-range system, such as an **Application System/400** (**AS/400**) or Unix system. The types of mainframes that you would most like to modernize will be either IBM or Unisys, and as you can see from this chapter, they have different approaches for how you modernize them.

We will get into the actual *how* to modernize later in the book. For now, the intent is to show that IBM and Unisys mainframes, while somewhat similar, are different systems that need to be approached differently.

We have now covered both common types of mainframes. The next chapter will cover midrange systems, such as the iSeries (AS/400) and **Advanced Interactive eXecutive** (**AIX**).

3
Midrange to Azure

In the world of legacy modernization, there are two tiers of computers that you will encounter. Mainframes, both IBM and Unisys as we discussed in the previous chapter, were typically the core mission-critical systems. The other systems we need to consider are midrange systems. Midrange systems often supported the mainframes and were very popular in branch banking and retail store environments. They also were the core systems for smaller businesses and agencies. In this chapter, we will cover the following topics:

- What is a midrange system?
- IBM iSeries
- IBM AIX

What is a midrange system?

For this book, we consider **IBM POWER**-based systems to be midrange systems. There are other legacy systems such as **Solaris** and **HP-UX** that some may also consider to be midrange systems. However, we will cover those in *Chapter 4*.

POWER systems

IBM POWER systems use a high-performance computer chip and are the latest generation of big-endian hardware that power the **iSeries** (also known as **AS/400**) and AIX (IBM's Unix variant) operating systems. POWER systems share some characteristics with the IBM zSeries hardware, such as the ability to be partitioned in **logical partitions** (**LPARs**), run big-endian bit ordering, and use **Extended Binary Coded Decimal Interchange Code** (**EBCDIC**) encoding. A single POWER platform can run LPARs with both iSeries and AIX. Finally, it's important to understand that applications developed for POWER operating systems are not compatible with x86-based hardware. This means that some type of transformation will be necessary if you want to move your applications to x86-based operating systems.

POWER-based hardware has the additional option of migrating to Azure, which includes running on POWER-based racks that are a part of selected Azure regions. This means you can move an on-premises LPAR to an LPAR running on supported versions of the iSeries and AIX operating systems on racks in Azure without changing the code or databases. This is true of applications that run on both iSeries and AIX:

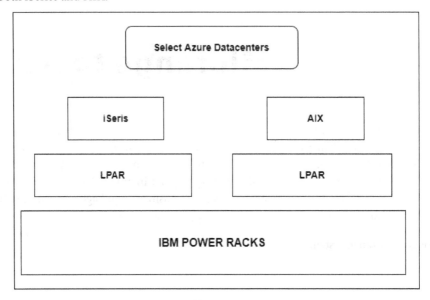

Figure 3.1 – POWER systems in Azure

In addition to lifting and shifting the LPARs to Azure, options exist to transform the code for both iSeries and AIX. This chapter will discuss migrating code from both iSeries and AIX in separate sections.

IBM iSeries

The IBM iSeries was developed in the mid-1980s to address a market for applications where a mainframe was either too large or too expensive. The original systems included IBM System 34/36/38 and the AS/400 systems. iSeries is often referred to as AS/400. These systems were cost-effective and efficient alternatives to the mainframe and required far less ongoing administration and maintenance.

OS/400 overview

While AS/400 refers to the hardware, **OS/400** refers to the operating system for iSeries. OS/400 is an operating system where the data, applications, scripting, and communications work tightly together. This makes OS/400 very efficient, but this also can make it difficult to migrate the applications without transforming the entire environment (apps, data, administration).

OS/400 runs applications written in both **Cobol** and **RPG**, with RPG having the lion's share (80%) of the applications. OS/400 uses a scripting language called **Control Language** (**CL**), which is tightly integrated with the application code and can actually be embedded with it. The iSeries also has an integrated development environment called **data description specification** (**DDS**), which is used to create screens that are tightly coupled to the data files.

To summarize, OS/400 is a tightly integrated operating system environment:

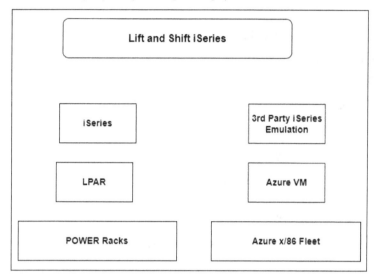

Figure 3.2 – Lift-and-shift options

However, that level of integration also makes applications written for OS/400 difficult to migrate. For this reason, options such as lifting and shifting LPARs and running emulation software are often considered.

Applications

In addition to applications developed in-house, many companies use packaged software from **independent software vendors** (**ISVs**) to run on the iSeries platform. Additionally, due to the nature of iSeries systems running on POWER, these systems can be readily deployed in remote locations such as stores, branch offices, or manufacturing plants. This has led to a rich ecosystem of applications designed for the iSeries. Some of the major types of applications include the following:

- **Enterprise resource planning** (**ERP**) systems
- Logistic systems
- Branch automation
- **Point of sale** (**POS**) systems

As touched upon previously in this chapter, options exist for modernizing POWER-based applications to Azure in addition to the methods discussed in *Chapter 2* regarding mainframes. The available options are as follows:

- Lift and shift LPARs to POWER racks on Azure. This option is available in some but not all Azure regions.

- Use iSeries emulation with x86 VMs in Azure. This requires the source code to be available as this needs to be recompiled for x86. However, the applications can remain in RPG or Cobol and do not require source code changes.

- Use a partner solution that supports RPG or Cobol. This will likely require some source code changes.

- Use a partner solution that will refactor the RPG or Cobol into Java or C#.

- Rewrite the solution to be Azure native.

- For ISV solutions, see whether the SV has an x86 version of their application.

The third, fourth, and fifth options are similar to the options that are available for the mainframe. In this case, you will need to look at the functionality provided by the iSeries platform and map that to the x86 target. Examples include the following:

- **Application server**: Is there an IBM or third-party application server?

- **Database**: Does the application use **Db2** for iSeries or flat files?

- **Transactions**: How are transactions handled?

- **Scripting**: How much CL scripting does the application have?

- **Interfaces**: What type of interfaces does the application need?

The following diagram shows the mapping of these services to Azure:

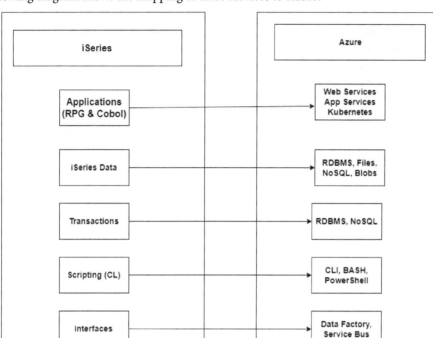

Figure 3.3 – iSeries to Azure service mappings

Another factor to consider with iSeries applications is that the tightly coupled nature of iSeries services makes these applications mostly monolithic and scale-up in nature. That does not mean that iSeries applications are not appropriate for remote (branch office) implementations. It does mean that any particular application is usually implemented on a single instance that you scale up the LPAR. For this reason, sizing the target platform on Azure is an important step to provide compute that matches the source implementation. This is true for either lift-and-shift cases to POWER on Azure or for computing equivalent performance if the application is rewritten for x86.

Data

The iSeries platform provides an integrated data system that includes relational (Db2 or i), logical, and physical files (sequential, indexed, or relative). These types of files on the iSeries are actually stored in the same physical database as Db2 but they usually have fixed-length records with fields, although they can be of variable length.

IBM has a version of Db2 specifically for the iSeries. This means that even if you wish to retain Db2 on a migration to x86 in Azure, a transformation is required for this to happen. For this reason, other relational databases are also considered if an application is rewritten for x86. Also, this makes the lift-and-shift or emulation options for moving to Azure attractive to some customers. If you use that route, the database schema and access remain intact. The following diagram depicts the option to transform or keep the data intact:

Figure 3.4 – iSeries to Azure data mappings

To summarize, the options for converting iSeries data are as follows:

- iSeries Db2
- Physical and logical flat files:
  - Use lift and shift or emulation to keep Db2 schema/SQL the same
  - Transform iSeries Db2 to a relational database that runs on Azure with either **platform as a service (PaaS)** or **infrastructure as a service (IaaS)**
- iSeries flat file with the **integrated file system (IFS)**:
  - Use lift-and-shift or emulation option to keep file access the same using Azure storage
  - Use third-party **Indexed Sequential Access Method** (**ISAM**) to retain flat file type access with code converted to x86
  - Convert flat files to a relational or NoSQL database with corresponding code and schema changes

Control Language

iSeries systems have a tightly integrated scripting system called **Control Language** (**CL**). CL can be traced back to IBM System 38 and can be used for running jobs and being embedded directly in the source code. It is common for RPG and COBOL programs on the iSeries to contain CL. For this reason, CL can be thought of as both a scripting language and an extension to programming source code.

CL can be handled in several different ways:

- Leave CL *as is* and use a lift-and-shift or emulation option
- Use third-party tools to convert the CL as part of code migration to Azure
- Rewrite the CL to a scripting language that runs on Azure such as **Bourne Again Shell** (**Bash**), **PowerShell**, or **Azure Cloud Shell command-line interface** (**CLI**)

If you choose the option to rewrite, be sure to address any CL that might be embedded in programs.

ISV versus homegrown solutions

As mentioned previously, the iSeries platform was used to develop homegrown applications but also has a rich ecosystem of ISVs developing **commercial off-the-shelf** (**COTS**) systems. There are fundamentally separate approaches for each type of system.

Homegrown (developed in-house) systems usually (but not always) have source code. This opens the door for the emulation option that needs to recompile source code in addition to the migrate option using third-party tools. The lift and shift of LPARs to POWER racks in Azure is also available.

ISV applications typically do not include source code. So, that eliminates the option of using emulation with recompile or third-party migration tools. However, ISVs that develop systems for iSeries may also have x86 versions of their products. You will need to contact the appropriate ISV to see whether any of the following is true:

- The ISV has an x86 version
- There is a migration path to the ISV version
- Any licensing requirement exists to migrate to the x86 version

If the ISV does not have an x86 version, the lift and shift to POWER on Azure, or the option to find a different ISV that does provide a COTS package that will run on Azure, is probably the best option. Many customers are open to alternatives for COTS packages for their current iSeries estate.

Administration

iSeries platforms can often run nearly autonomously in remote environments. However, they also require administrative tasks such as the following:

- Backup and recovery
- Software updates
- Monitoring and alerts
- Daily job submission

These are normal administrative tasks that can typically be automated in Azure to keep the autonomous nature of the iSeries applications you want to move to Azure.

Additionally, many iSeries systems need to integrate with other systems in a customer's estate. This is especially true for remote implementations such as branch office systems that need to integrate with the corporate systems at the home office. The types of integration include the following:

- **Database integration**: Typically using **Open Database Connectivity (ODBC)** for Db2 on the iSeries
- **File system sharing**: Typically using **Network File System (NFS)**
- **Network communication**: Typically using **System Network Architecture (SNA)** with **Logical Unit 6.2 (LU.6.2)** over **Internet Protocol (IP)**

These interfaces will likely need to be addressed or updated using newer methods or protocols. Azure provides similar functionality, as shown in the following diagram:

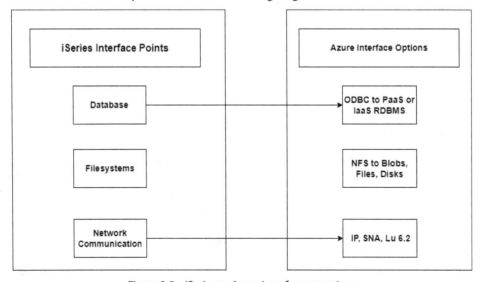

Figure 3.5 – iSeries to Azure interface mappings

To summarize, there are well-defined methodologies and patterns to migrate iSeries applications to Azure. However, the proper pattern to use will depend on your specific situation and the applications you use.

IBM AIX

AIX was released by IBM in 1986 and stands for **Advanced Interactive eXecutive**. It was their proprietary offering of UNIX that was based on the UNIX System V. Over the years, they ported it to a variety of platforms including the early RISC-based RS/6000, their System 370 Mainframes, PS-2 Personal Computers (remember them?), and more recently, the platform you will more than likely find it on, the Power Platform, which is also what the IBM i Series (AS/400) runs on. This often creates confusion when considering your approach to migration because there are two distinct operating systems that IBM offers on the same hardware platform, and the considerations for migrating the applications that run on them are very different. Let's take a deeper look at AIX and the special considerations it requires for migration to the distributed x86 platform.

AIX overview

If you are familiar with UNIX (or Linux), especially BSD UNIX, AIX will seem very familiar. Except for a few considerations that are mainly centered around networking, AIX is BSD-compatible. As far as scripting shells for AIX go, the default is the Korn shell, but there are others that are available too so your choice is not limited. Your favorite UNIX tools are available as well. The filesystems currently supported are JFS, JFS2, ISO 9660, UDF, NFS, SMBFS, and GPFS.

The main difference between AIX and the other flavors of UNIX and Linux that you might be familiar with is that it only runs on the IBM Power Platform offered by IBM. This includes POWER, PowerPC-AS, PowerPC, and Power ISA. This is a pretty big difference when it comes to porting or migrating applications because there are some substantial differences between x86-based and Power-based architectures. Let's dive deeper into that subject.

AIX versus Linux — what's the difference?

AIX and Linux are both based on the UNIX operating system, and their approaches to scripting and programming are very comparable to one another. Nevertheless, in order for an application to continue functioning correctly after the migration, it is necessary to resolve the variations in syntax and directory structure. The fact that AIX is based on the **reduced instruction set computing** (**RISC**) processors (that is, IBM Power PC, IBM z-series, and Sun SPARC), while the majority of Linux systems are based on the Intel x86 CPU, is by far the most significant distinction between the two operating systems. RISC processors were a type of microprocessor that was built with the intention of concentrating on a more condensed and efficient set of instructions as opposed to a wider and more diverse set of instructions.

Endianness is the term that is used to describe how a system stores bytes in computer memory. RISC processors and x86 processors have different endiannesses. Large endian systems are computers that store the most significant value (large value) first, which means they store it in the lowest storage address. This is why RISC-based computers are known as big-endian systems. The x86 microprocessor powers the majority of computers running Linux:

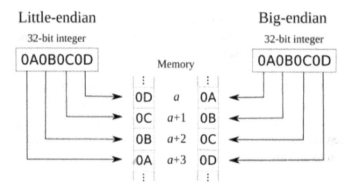

Figure 3.6 – How little-endian and big-endian bit encoding are represented in computer memory

As you can see from the preceding graphic, big-endian might seem more intuitive in that it aligns with the way we typically process information (i.e., from left to right), but little-endian has a benefit when it comes to adding to numbers in memory. In the end (pun intended), the decision is arbitrary.

So, what does all of this mean? It means that the executables from each platform are different and cannot be ported from one platform to the other. The bottom line is that all of the source code you are considering for migration will need to be compiled for the target platform. The exception to this is Java application code, which brings us to the topic of applications.

Applications

In the AIX application world, there are basically two types of applications you will encounter when you are considering a migration. The first category we will call **native applications**. These are applications that compile down to an executable that interoperates with the microprocessor platform and is dependent on that platform. The code executes faster, but it is tied to that platform.

The second type of applications that you will typically find on an AIX system are **Java applications**. Actually, I have found these to be more prevalent. So, if you were a developer back in the 1990s as I was, you might recall that the reason Java was all the rage back then, and to a certain extent still is today, is because Java offered platform independence by providing an application execution environment that was abstracted from the operating system. This meant that your Java code was not tied to the platform it was running on. What this allows is the ability to port your Java application byte code (the Java executables) from one platform to the other and it will usually work—provided the Java platforms are compatible.

I myself experienced this first hand when I was involved in an AIX to Linux migration of a major statewide health and human services system. We were migrating a system from WAS WebSphere Application Server on the AIX side to Apache on the Azure/Linux side. Aside from some minor adjustments, the ported code worked well.

Both of the approaches that we have talked about assume one thing: that you legally and physically have the source code. If you do not, then you are probably looking at a rewrite or rearchitecting of the workloads.

Azure Spring Cloud

Every strategy that we have discussed up to this point is predicated on moving AIX to a virtual machine (VM) hosted by Azure, which normally runs Linux. Azure Spring Cloud is yet another alternative that was made accessible to customers in October 2019. With Azure Spring Cloud, you are able to run apps written in Java within a cloud-native Java environment with the help of this completely managed service. You will then be able to advance beyond a conventional VM deployment and into a microservices architecture that is more flexible and agile.

Data

So, from an application standpoint, there are two main approaches to migrating your legacy applications to Azure/Linux. The other tier we have to migrate is the database tier. Many AIX systems use Db2 as their database since it is the premier database offered by IBM. Having said that, you could also encounter Oracle, IMS DB, Informix, or a variety of other databases. For the purposes of this book, we will focus on the main databases you might have to deal with.

First up is Db2. As with the others that we will consider, you have options. If you have an investment in people who are familiar with Db2 (and not other database platforms), you might very well want to port the Db2 data from Db2 AIX to Db2 LUW (Linux, UNIX, Windows) on Azure. The consideration for this would be the personnel maintaining the database and the investment in stored procedures. I personally would recommend Azure SQL MI (where **MI** stands for **managed instance**) since it is a **database as a service** (**DaaS**) offering in Azure that has almost all of the benefits of SQL Server Enterprise IaaS and is an enterprise-class database offering.

Another database that is very popular is Oracle. For Oracle, the same methodology applies. You can port the Oracle databases from AIX to Oracle on Windows or Linux in Azure. Again, I would personally recommend Azure SQL MI. However, lately, I have seen a trend toward Azure PostgreSQL for Oracle migrations. There are cost savings with this approach, but there is also the reliance on some open source solutions for certain functionality. If you are okay with that, this might be the right decision for you.

For other databases, the options will vary. A good example is Informix. There is no Informix equivalent in Azure, so that database will need to be migrated to another. Typically, just like Db2 and Oracle, Azure SQL MI is a great choice.

Before we leave this topic, it is worth bringing up licensing. IBM and Oracle licensing can vary significantly—in a bad way (i.e., more expensive)—when you migrate, say, from Db2 AIX to Db2 LUW on Azure. Just something to be aware of.

COTS applications

Another option for migrating your application workloads to Azure is what is called COTS to COTS. A good example of this is the very well-known **System Analysis Program (SAP)**, which I think you are probably familiar with if you are in the information technology arena. The approach for this is much easier than the options we have been considering so far.

Here, what is necessary is to find the equivalent x86 offering of the COTS solution you are using and port the data from one platform to the other. I have seen quite a few inventory management solutions and content management solutions that are good candidates for this approach.

In fact, the IBM **Global Solutions Directory** can help out here. It offers a very substantial list of solutions available on multiple platforms. The URL at the time of writing is `https://www.ibm.com/partnerworld/bpdirectory/`. Keep in mind that URLs and the contents they reference change over time. Rather than provide the list here, we suggest that you reference the IBM site for the most complete and up-to-date list.

This can be very beneficial in helping you decide the approach to take.

Administration

From an administration standpoint, since you are migrating from one UNIX variant to another UNIX variant, there should not be a significant change. However, something will change and as we all know, *the devil is in the details*. There will be some training required to maintain the operating systems and application, but typically, the automation and ease of administration in the Azure cloud are drivers for why customers want to do this in the first place.

So, if you are planning on migrating from an IBM AIX system, this chapter has hopefully provided you with some guidance and cost-effective approaches. Keep in mind that every UNIX variant, of which AIX is one, can have a different path for modernizing and migrating. You will need to pick the path that best matches your desired outcome.

Summary

This chapter has covered the options for migrating from IBM iSeries and AIX systems. What these systems have in common is that they both run on the IBM POWER platform. However, as you can see, for midrange systems, there are a few very different approaches to migration. You will need to consider what your end state looks like and drive toward that with the options presented here. In this book, we have mostly focused on the technical aspects of migration. As mentioned earlier, licensing also needs to be considered.

We've now looked at midrange systems. Next, we will look at enterprise UNIX systems.

4

Modernizing Legacy UNIX Systems

UNIX is an operating environment that is still in use by a number of vendors. This chapter will look at the UNIX variants and the options for modernizing to Azure.

In this chapter, we will cover the following topics:

- Current UNIX landscape
- Migrating Solaris
- Migrating HP-UX
- Other UNIX variants

The current UNIX landscape

The UNIX system was originally developed by Bell Labs over 40 years ago as the first **Open Systems** operating system. The intent of Bell Labs was to have an underlying operating system kernel and shell that was not tied to specific hardware. To accomplish this, Bell Labs created underlying operating system capabilities and provided hooks to allow specific vendors to adapt UNIX to their hardware. This resulted in similar, but not identical, variants of UNIX offered by the various vendors. This means there is no binary compatibility for applications written on different variants. However, users should be able to recompile applications for a specific variant without a lot of changes. In reality, this is sometimes the case, but not always.

The result is several variants of UNIX that have similar, but not identical, paths to modernize to Azure. The primary variants of UNIX this book will discuss are as follows:

- **AIX**: Covered in *Chapter 3*
- **Solaris**: Covered in this chapter
- **HP-UX**: Covered in this chapter

We will also discuss options for migrating to Azure for variants not covered in *Chapter 3* and this chapter.

BSD System 4.3 versus System V

Most UNIX systems are based on the **Berkeley Software Distribution** (**BSD**), such as BSD extensions to AIX, and System V, such as Solaris and HP-UX. While BSD and System V are similar, there are differences in operating system commands and programming interfaces.

Old UNIX versus modern UNIX

Since UNIX has been around for a long time, it has run on various types of hardware. The most common systems available today run either on **Reduced Instruction Set Chip** (**RISC**) or x86 and Itanium-based hardware.

Older versions of both Solaris (SPARC) and HP-UX (PA-RISC) run on RISC hardware. Newer versions of Solaris are typically run on x86, with some that still run on RISC. Newer versions of HP-UX run on Itanium.

UNIX scripts

When modernizing UNIX environments, it is important to understand the role that UNIX shell scripts play. Shell scripts initiate executable binaries and can be mini-programs themselves. Common UNIX shell scripts are as follows:

- **Bourne**: Probably the most common
- **Korn**: A modernized version of Bourne
- **C**: Modeled after the C programming language, and often preferred by developers

Applications

Applications written for UNIX are typically compiled into binaries using a binary-compatible compiler, such as variant-specific C/C++ or Java.

Since UNIX has been around for a long time, there are many applications and ISV systems that may need to be addressed in a migration.

Now that we've looked at many of the ways customers use UNIX, along with the tools UNIX provides, we will look at some of the major versions of UNIX, including Solaris and HP-UX.

Migrating Solaris

Solaris was developed as a variant of System V UNIX by Sun Microsystems and was originally powered by a proprietary RISC-based CPU known as SPARC. Later versions of Solaris became available with x86-based computer chips.

Solaris enjoyed widespread adoption in the mid to late 1990s, and powered many of the `.com` solutions in the late 1990s and early 2000s. Sun was later acquired by Oracle after the `.com` boom ended. Solaris continues to be available for both SPARC and x86-based CPUs.

Solaris was widely used for the following:

- Web servers

- Relational databases

- Enterprise applications, such as SAP

- **Lightweight Directory Access Protocol (LDAP)** servers

Solaris was considered the most feature-rich variant of UNIX and was deployed in both commercial and government applications.

SPARC versus x86

SPARC versions of Solaris are typically older. But they are still in use today, especially for applications that require high-clock-speed CPUs. These are typically scale-up-type architectures, such as relational databases, or any other application that relies on CPU clock speed for performance.

The x86 version of Solaris gained popularity in the mid-2000s and is the most common version in use today. There is a wide variety of solutions available for x86 Solaris, and x86 Solaris is supported by many ISVs. These include relational databases, web servers, and enterprise application servers.

Emulation options

The quickest way to move Solaris to Azure is to use a third-party emulation solution that runs on IaaS VMs in Azure. This option does not require recompilation and is available for both SPARC and x86 versions of Solaris. Separate licensing is required for the third-party emulation and this is not offered by Microsoft for Azure.

The following diagram shows how UNIX emulation from Solaris and HP-UX PA-RISC can be hosted in Azure on VMs:

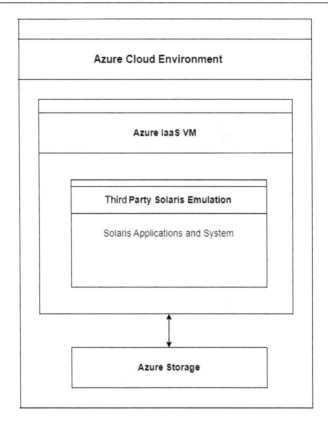

Figure 4.1 – Solaris emulation diagram

The main issues for using Solaris emulation on Azure deal with sizing for both compute and storage. There is some overhead required for the third-party emulation software that needs to be accounted for. However, modern VMs available in Azure will often provide more compute power than the legacy CPUs used by existing implementations of Solaris.

Converting to Linux

The most common way to convert from Solaris to Azure is to convert to Linux and host the Linux VMs in Azure using IaaS. This can be done using a number of different methods:

- **Recompile applications to target x86 Linux**: This requires the source code and will require changing the scripts from Solaris-based to Linux-based.

- **Using a Linux version of ISV software**: Many ISVs that provide Solaris solutions also provide solutions for Linux. This might require an upgrade to a more current version of the ISV software. Additionally, there may be some surrounding applications that will need to be converted as described in the previous bullet, and scripts will likely require conversion.

- **Converting to a new software package specifically designed for Linux**: ISVs typically provide migration paths and automated tools to assist with this type of migration.

The following diagram shows how operating system services on Solaris can map to similar services on Linux IaaS VMs:

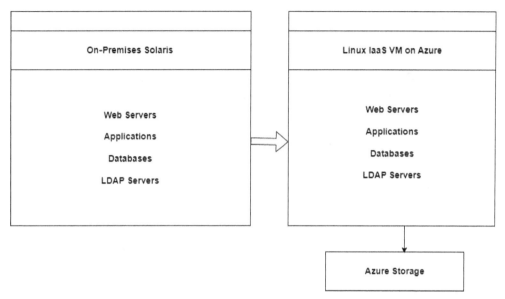

Figure 4.2 – Solaris to Linux diagram

Converting Java-based systems from Solaris to Linux is typically much easier than converting C/C++ systems, and C/C++ systems often depend on libraries and services unique to Solaris.

Once on Linux, the target VMs can be hosted on Azure using IaaS. As with emulation, VM and storage sizing are important factors to consider.

Converting to PaaS services

Since many of the applications originally developed for Solaris used services such as web services and relational databases, you might want to see whether PaaS services on Azure are an option. The types of services Azure PaaS can offer include the following:

- Web services
- App services
- PaaS databases
- Eventing and queuing

The following diagram shows how applications and operating system services on Solaris can map to cloud-native PaaS services on Azure:

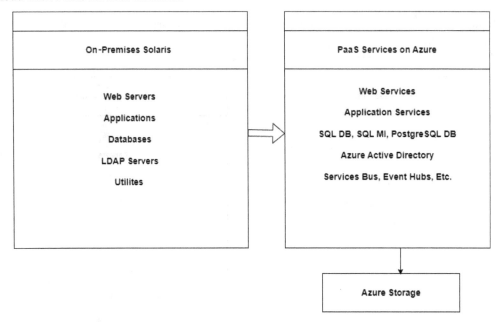

Figure 4.3 – Solaris to Azure PaaS diagram

PaaS services are typically desirable since they require less maintenance once converted to Azure.

Hosting in Azure

Hosting in Azure allows a number of cloud services to be available for the system or application you are moving to Azure. These include the following:

- Virtual private networks
- Security options: AAD, security groups, and so on
- High-availability options: Both within and across regions
- Backup and disaster recovery options
- Load balancing
- On-demand scalability

Now that we've looked at moving Solaris to Azure, we will next look at moving HP-UX to Azure.

Migrating HP-UX

This section may seem to somewhat repeat what was discussed in the previous section on Solaris. That's because both HP-UX and Solaris were developed as a variant of System V UNIX. HP-UX, however, was developed by **Hewlett-Packard** (**HP**). As with Solaris, the first versions of HP-UX were powered by RISC cups. This was known as the PA-RISC version. Later versions of HP-UX were powered by Intel Itanium CPUs.

HP-UX had many implementations in the mid-to-late 1990s, which are still in use today. The HP-UX is now developed by **Hewlett-Packard Enterprise** (**HPE**).

HP-UX was widely used for the following:

- Web servers
- Relational database
- Enterprise application

HP-UX was considered one of the most performant and highly available variants of UNIX and was deployed in commercial **independent software vendor** (**ISV**) offerings.

PA-RISC versus Itanium

PA-RISC was discontinued. But there are still customers that use this version. The Itanium version is used by the majority of customers today and continues to have new releases.

Emulation options

The quickest way to move PA-RISC HP-UX to Azure is to use a third-party emulation solution that runs on IaaS VMs in Azure. Separate licensing is required for the third-party emulation, and this is not offered by Microsoft for Azure.

The following diagram shows how HP-UX PA-RISC can use third-party emulation on Azure:

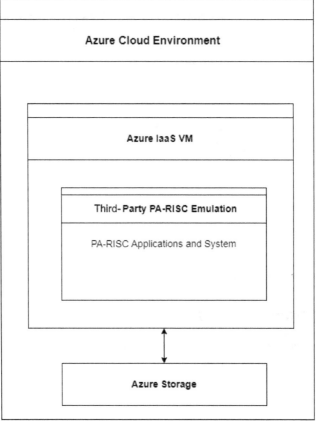

Figure 4.4 – PA-RISC emulation diagram

The main issues for using PA-RISC emulation on Azure deal with sizing for both compute and storage. There is some overhead required for the third-party emulation software that needs to be accounted for. However, modern VMs available in Azure will often provide more compute power than the legacy CPUs used by existing implementations of PA-RISC.

Converting to Linux

The most common way to convert from HP-UX to Azure is to convert to Linux and host the Linux VMs in Azure using IaaS. This can be done using a number of different methods:

- **Recompile applications to target x86 Linux**: This requires the source code and will require changing the scripts from HP-UX-based to Linux-based.

- **Using a Linux version of ISV software**: Many ISVs that provide HP-UX solutions also provide solutions for Linux. This might require an upgrade to a more current version of the ISV software. Additionally, there may be some surrounding applications that will need to be converted as described in the previous bullet, and scripts will likely require conversion.

- **Converting to a new software package specifically designed for Linux**: ISVs typically provide migration paths and automated tools to assist with this type of migration.

The following diagram shows how HP-UX applications and services can map to services on Linux IaaS VMs in Azure:

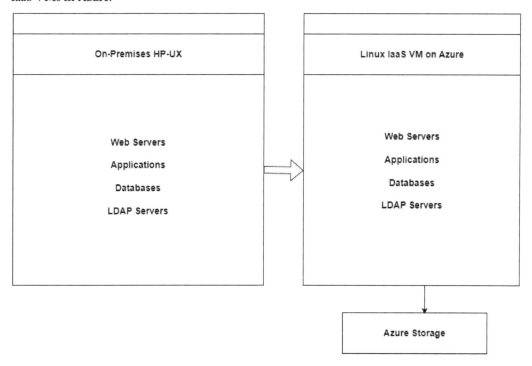

Figure 4.5 – HP-UX to Linux diagram

Converting Java-based systems from HP-UX to Linux is typically much easier than converting C/C++ systems, and C/C++ systems often depend on libraries and services unique to HP-UX.

Once on Linux, the target VMs can be hosted on Azure using IaaS. As with emulation, VM and storage sizing are important factors to consider.

Converting to Azure PaaS

Since many of the applications were originally developed for HP-UX using services such as web services and relational databases, you might want to see whether PaaS services on Azure are an option. The types of services Azure PaaS can offer include the following:

- Web services

- App services

- PaaS databases

- Eventing and queuing

The following diagram shows how HP-UX applications and services can be used in Azure native PaaS services:

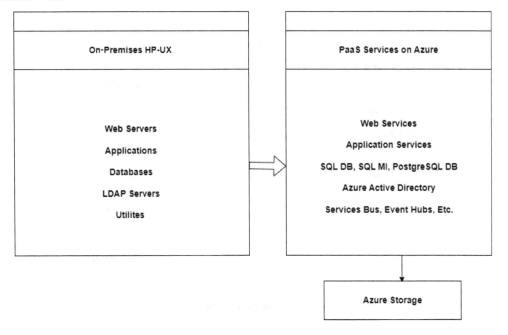

Figure 4.6 – HP-UX to Azure PaaS diagram

PaaS services are typically desirable since they require less maintenance once converted to Azure.

Hosting in Azure

Hosting in Azure allows a number of cloud services to be available for the system or application you are moving to Azure. These include the following:

- Virtual private networks
- Security options: AAD, security groups, and so on
- High-availability options: Both within and across regions
- Backup and disaster recovery options
- Load balancing
- On-demand scalability

We've now looked at HP-UX and how to migrate HP-UX to Azure. We will briefly look at what should be done with other UNIX variants next.

Other UNIX variants

You might find other variants of both BSD 4.3 and System V UNIX. In this case, you will need to follow the option for recompiling as emulation options are not available.

Summary

There are existing patterns for converting UNIX variants to Azure. The main decision you will need to make is what type of target architecture you want on Azure. This will often be determined by how you want to maintain the application moving forward. If you want to keep this pretty much the same in the Azure version, then an emulation option (if available) or recompilation to VMs might be the right way to go. If you want to have a more cloud-native solution, then using PaaS services should be investigated as an option. Finally, if you are moving an ISV application from UNIX to Azure, you need to move forward with the vendor-recommended solution.

We've looked at moving from various types of legacy environments in the first four chapters, including this one on UNIX. We will now look at Azure services that enterprise applications can use in the next two chapters.

Part 2:
Architecture Options

This part will look at the architecture options for the estate in Azure in which your migrated systems will land.

This part contains the following chapters:

- *Chapter 5, An Overview of the Microsoft Azure Cloud Platform*
- *Chapter 6, Azure Cloud Architecture Overview for Mission-Critical Workloads*
- *Chapter 7, Behold the Monolith – Pros and Cons of a Monolithic Architecture*

5

An Overview of the Microsoft Azure Cloud Platform

The goal of this book is to guide you on your legacy modernization journey to the Microsoft Azure cloud. Up to now, we have spent some time providing an overview of the legacy platforms you will probably encounter. Now, it's time to take a look at the *landing zone*, which is the Microsoft Azure platform itself. Here, in *Chapter 5*, we will provide an overview of the Azure platform, with a specific focus on **Infrastructure as a Service (IaaS)** and what it means to be *cloud native*. So, let's get started.

However, before we dive into a brief history of Azure, we should define some terms that we will be using for clarity. We list them here:

- **IaaS**: With IaaS, you essentially have a **virtual machine (VM)** on demand on a pay-as-you-go basis. With IaaS, the customer is responsible for the updating, administration, and management of the operating system. VMs come in different configurations, and the reason customers like this offering is that it allows you to easily migrate your existing x86-based VMs to Azure.

- **Platform as a Service (PaaS)**: PaaS offers an application *platform* for applications. The benefit of this is operating system responsibilities are shielded from you, and you only need to focus on the data, application, and middleware necessary for supporting the application.

- **Software as a Service (SaaS)**: SaaS is a software offering. Azure allows you to create such an offering, whereby you create a solution that runs in the Azure platform and offer that solution to customers. Good examples are Office 365 and Salesforce.

The following screenshot provides a visual representation of IaaS, PaaS, and SaaS:

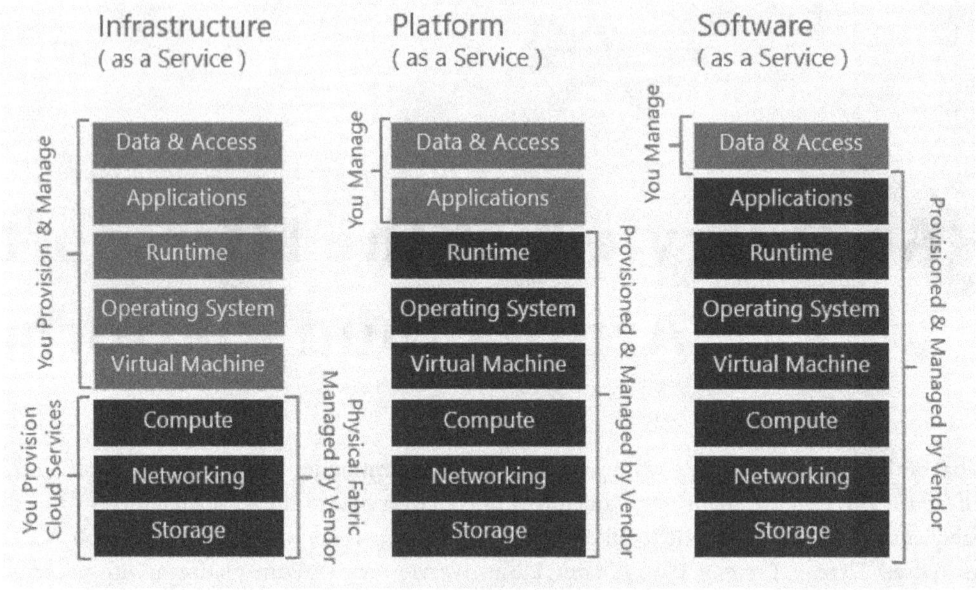

Figure 5.1 – IaaS versus PaaS versus SaaS

After an overview of the Azure platform, we will provide some sample configurations of various deployments for mainframe and legacy workloads. These are based on best practices and current offerings at the time of this publication.

This chapter covers the following topics:

- A brief history of Azure
- Azure Regions
- Azure Stack
- Azure compute
- Azure storage
- Azure networking
- Azure databases
- Azure security and identity

- Understanding the Azure approach to hybrid

- Deploying and maintaining systems on Azure

- Looking at cloud native, serverless computing, and microservices

A brief history of Azure

I was actually working at Microsoft at the time of Azure's launch. Internally, the project code name was *Project Red Dog*, supposedly named after a bar in Silicon Valley. It was launched in October 2008 at the **Professional Developers Conference** (**PDC**) and had **General Availability** (**GA**) to the public in February 2010. It was initially solely based on Windows and was actually called the *Windows Azure platform*. I still have a sticker of the original logo. In 2014, the "Windows" part was dropped from the name since it later adopted the Linux operating system as well, and it was called *Microsoft Azure*. As a side note, over 50% of the VMs that run in Azure are Linux as of the publishing of this book.

If you were not working or tracking Azure when it came out, you might find it very interesting that when it was first released, it was only a PaaS offering with no provision for IaaS. It was conceptually ahead of its time. The primary assets you had to deal with were *web roles* and *worker roles*, each designed for a specific purpose (that is, web tier and application tier specifically). This meant that it was primarily being targeted toward web application developers and web-based applications. However, as it turned out, customers also wanted IaaS so that they could run already existing on-premises workloads in Azure. This was added later in 2014 when Azure switched to a completely new model based on something called **Azure Resource Manager** (**ARM**). To this day, Azure is still based on this much more flexible model and supports IaaS and cloud-native deployments.

Now that we have some context about how Azure came to be a world-class cloud platform, let's look at how it is deployed across the world.

Azure Regions

At the time of writing this, there are 60+ Azure Regions with 160+ datacenters located in 140 countries. If there is not yet an Azure Region in your country or locale, you might want to inquire with Microsoft, as there could be one coming soon. This makes Microsoft Azure one of the largest public cloud providers. Along with this, there are also *Azure Sovereign Clouds*, which adhere to specific laws and guidelines for specific countries.

The following screenshot shows the current deployments of Azure Regions across the world as of the publishing of this book:

Figure 5.2 – Azure Regions' global footprint circa 2021

So, you might ask: *What exactly is the difference between an Azure Region and an Azure datacenter?* It might be good to define some terms to provide some clarity on this topic, so let's do that now:

- **Azure Region**: An Azure Region is really just a collection of Azure datacenters that are located in a specific Region. They are connected by a high-speed *latency-defined perimeter*, which means that the network latency must meet a certain threshold to allow for communication between datacenters.

- **Azure datacenter**: In Azure speak, an Azure datacenter is what is also known as an **Availability Zone**. Availability Zones, as the name implies, provide tolerance to failures by providing redundancy and logical isolation of Azure services for the Region. **Azure Availability Zones** are unique physical locations within an Azure Region and offer **high availability (HA)** to protect your applications and data from datacenter failures. Each zone is made up of one or more data centers equipped with independent power, cooling, and networking. This resiliency to failure is provided by a minimum of three Azure Availability Zones per Region. They are connected by a high-speed network that must have a round-trip latency of no more than 2 ms. This allows Azure to keep VMs, data, and applications highly available—five nines of availability, in fact, which is extremely good. It should be mentioned here that in order for your application to be resilient and highly available, you must leverage the Availability Zones in your deployment. I mention this because it is possible to not leverage them. A good example is if you deploy only one VM and do not include it in an Availability Zone, you will not reap the benefit the Availability Zones provide. It is very easy to do this. Essentially, you just need to turn them on.

To find out the locality of Azure Regions near you, you should check out this site: `https://azure.microsoft.com/en-us/explore/global-infrastructure/geographies/#overview`.

Keep in mind that it is constantly being updated. Another point to bring up in this topic is that not all Azure Regions are the same. They are different sizes, and also, not every Azure Region has all of the possible available Azure services. This can be very important for a deployment project you are working on. You can check the availability of Azure services by Region at the following page (`https://azure.microsoft.com/en-us/explore/global-infrastructure/products-by-region/`). We have looked at the current worldwide deployment of Azure Regions. Now, let's look at a more local implementation of Azure—Azure Stack.

Azure Stack

If there is no Azure Region near you, or perhaps for technical or compliance reasons you want to have a local implementation of Azure, you need to look into Azure Stack. Azure Stack is a local implementation of Azure that you can deploy in your own data center. Keep in mind, however, that the services available on Azure Stack are a subset of the services available in the public cloud offering. Having said that, you can leverage the majority of the IaaS and PaaS services. This might be a good solution if you want to minimize latency between applications running in Azure and your mainframe installation.

Azure Stack comes in three separate offerings, as follows:

- **Azure Stack Edge**: Azure Stack Edge is designed primarily for **Internet of Things** (**IoT**) and edge AI-type workloads. It employs hardware-accelerated **machine learning** (**ML**) for fast analysis at edge locations.

- **Azure Stack HCI**: This offering provides a hyperconverged hybrid solution that is integrated into Azure. It is primarily targeted toward branch and high-performance workloads.

- **Azure Stack Hub**: This allows you to have a subset of Azure in your on-premises data center. It can be used for application modernization and connected and disconnected scenarios.

So, there you have the current deployments of Azure, both local and worldwide. Now, let's look at the components that make up the Azure cloud platform.

Azure compute

When we talk about Azure compute, we need to look at and consider the various forms and offerings in which Azure makes compute available. At its most basic level, compute refers to the CPU capacity available for running your applications. It starts to get interesting when we find that we need to leverage cloud features such as the ability to scale on demand. Or, do we want to move away from standard VMs and more toward Docker containers, offered in **Azure Kubernetes Service** (**AKS**) or **Azure Container Instances** (**ACI**)?

These are all extremely important questions that you will need to consider when you are planning your migration.

The best way to do this is to provide a quick rundown on the types of compute offerings that are available in Azure, but before we do that, let's talk about the types of compute CPUs available. It would be easy to get into the weeds on this topic, so I will keep it brief. Essentially, Azure runs many different forms of x86 vCPUs. When we state **vCPU**, it means **virtual CPU**, which is what the Azure offering is based on. There are two high-level types of x86 vCPUs in Azure—those based on Intel and those based on AMD. Each offering has different performance characteristics that you will need to consider, and exploring them is beyond the scope of this book. Another thing to keep in mind is that the **Azure hypervisor**, which is the system that allows the magic of virtualization within Azure, is based on **Microsoft Hyper-V**. Hyper-V is a virtualization technology that was first introduced in Windows Server 2008 and is used for many on-premises data centers.

I should also mention that other *non-x86*-based CPUs are available as well with something called Azure Dedicated Host. A good example of this is **Skytap** on Azure—a specialized implementation of the IBM Power platform, which we looked at in *Chapter 3*.

Now, let's take a quick look at the types of Azure compute offerings available:

- **VMs**: Specific preconfigured Linux and Windows VMs with various configurations. You can import your predefined Hyper-V VM configurations as well. If your on-premises data center leverages VMware virtualization, you can migrate this to Azure as well using the Azure Migrate tool, which handles the conversion from VMware ESXi or NSX implementations.

- **VM Scale Sets**: Provide HA and scalability by allowing you to create thousands of VMs in minutes and deprovision them as well.

- **Azure Spot VMs**: Can provide deep discounts for leveraging unused capacity in Azure.

- **AKS**: A managed Kubernetes service for running your container-based applications.

- **Azure Functions**: Event-driven functions in a serverless architecture.

- **Azure Service Fabric**: For running and orchestrating microservices.

- **Azure App Service**: Quickly create cloud applications for web and mobile.

- **ACI**: ACI simplifies the running of container applications. Some people find the management and administration of Kubernetes and AKS too complex. ACI provides an easier implementation for deploying container-based applications.

- **Azure Batch**: Cloud-based job scheduling. Azure Batch is a managed service for running large-scale parallel and **high-performance computing** (**HPC**) workloads.

- **Azure Cloud Services**: A number of specialized services for running cloud applications and APIs.

- **Azure Dedicated Host**: Deploy your Azure VM on a physical server only used by your organization.

So, as you can see, there are quite a few services you can target for mainframe workloads. Here is a great article from Microsoft on which service might be appropriate for your workload: `https://learn.microsoft.com/en-us/azure/architecture/guide/technology-choices/compute-decision-tree`.

Now that we have explored the Azure approach for compute, let's turn our gaze to storage.

Azure storage

Storage in Azure is provided through seven separate data service offerings. Each one has a specific purpose and performance characteristic. The property common to all of them is that they are all offered through a storage account. Think of that as a high-level container for all of the offerings we are going to look at. What this means is that before you can create—for example—an Azure blob, you will first need to create a storage account. Let's take a look at the various offerings:

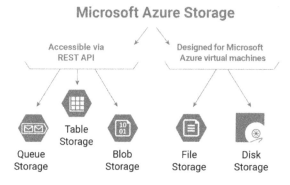

Figure 5.3 – Various Microsoft Azure storage offerings

We'll start with the ones that you will probably use most in your modernization journey:

- **Azure Disk Storage**: More popularly known as **Azure Managed Disks**. These are what you will use when you are deploying your VM or containers. There is quite a variety of **Stock-Keeping Units (SKUs)** available, and each one has its own performance characteristics.

- **Azure Files**: Azure Files offers fully managed file shares in Azure that are accessible via the **Server Message Block (SMB)** and **Network File System (NFS)** protocols and Azure Files REST APIs. They can be mounted from VMs in the cloud or on-premises and are accessible from both Linux and Windows VMs. They can also be synced with other files shared with the Azure File Sync service.

- **Azure Elastic SAN**: This is in preview at the time of writing this book. Azure Elastic SAN provides an integrated solution that simplifies the deployment and management of a **storage area network (SAN)**. It also offers built-in HA and is targeted toward large-scale I/O-intensive workloads such as relational databases.

- **Azure Queue Storage**: Primarily designed for message-based applications, Azure Queue Storage allows you to access messages from anywhere via HTTP or HTTPS. Usually used to create work backlogs for processing.

- **Azure Table Storage**: Provides a service that stores non-relational structured data in the cloud. This is also known as NoSQL. This is schemaless data, much as with mainframe **Virtual Storage Access Method** (**VSAM**) files. It implies that the application needs to know how to read and interpret the data. It is extremely cost-effective, but it should be noted that much of the functionality of Azure Table Storage can also be provided by Cosmos DB, which is a newer service in Azure.

- **Azure NetApp Files**: Azure NetApp Files is a fully provisioned implementation of NetApp files in Azure. It is extremely fast and provides shared file storage for **line-of-business** (**LOB**) applications.

So, there you have a quick rundown of the approach and types of storage available in Azure. We will look more closely at the database offerings available in Azure right after we take a look at the Azure approach to networking.

Azure networking

The Azure platform implements what is known as software-defined networking for the majority of the Azure fleet. This differs from Azure dedicated, where your workloads can actually be running on non-virtualized hardware. For the rest of this section, we will not address non-virtual implementations since you are probably already familiar with them.

An entire book could be devoted to Azure networking, but for the purposes of this book, we will focus on the key aspects that you need to be aware of on your modernization journey. They can broadly be thought of in two categories, as follows:

- Services that run in Azure

- Services that connect Azure to your on-premises data center

Let's start with those that are within Azure first.

Understanding internal Azure networking services

Let us take a look at Azure's internal networking services:

- **Azure Virtual Network**: An Azure Virtual Network is the most fundamental component of your **virtual private network** (**VPN**) in the Azure cloud. Most Azure architects refer to them as VNets. They allow for VMs and other Azure resources to communicate with each other inside the VNet, outside the VNet, and on-premises or in another public cloud. They are very similar to more traditional on-premise networks you might have worked on in the past, with a few added benefits such as scalability, availability, and isolation offered by being software-defined.

When we say that they are the most fundamental building block, consider this—you cannot create a VM within Azure unless it lives within a VNet. When you think about it, what would be the point of creating a VM if you could not communicate with it? Whether your migrated workload lives in a VM or an **AKS cluster** (**AKSC**), it will probably live inside a VNet.

You can create subnets within an Azure VNet. You can also filter traffic between the subnets, which is a best practice when you are deploying a three- or n-tier type of architecture for your migration workloads.

Following are some of the additional features of VNets:

- **Network security groups** (**NSGs**): The next fundamental building block to dive into is NSGs, which allow you to filter network traffic between Azure resources located within a VNet or outside as well. They are typically assigned to VMs or an AKSC. They function similarly to a firewall but are easier to implement. By effectively using NSGs, you can secure your migration workloads within the Azure cloud.

- **Azure Virtual Network peering**: Multiple VNets can be *peered* together in Azure, allowing them to appear as one VNet for connection purposes. All traffic between the peers is routed through the Azure network backbone, so it is extremely fast.

- **Azure Load Balancer**: Just as with a load balancer in your on-premises network, Azure Load Balancer balances inbound network traffic between a pair or configuration of Azure resources. It also allows you the routing to be based on the capacity of the resources it is routing to, which is very useful. It comes in two flavors, as follows:

  - **Public load balancer**: You would use a public load balancer to balance traffic coming in from the public internet (that is, public IP addresses) destined for internal resources.

  - **Internal (or private) load balancer**: You would use this to balance internal traffic coming from another Azure resource from either inside or outside (that is, in another VNet).

- **Azure Application Gateway**: When you need to do deeper filtering of incoming traffic at the application layer, you will need to leverage Azure Application Gateway. The key difference between this and the Azure Load Balancer service we just discussed is that load balancers such as Azure Load Balancer filter traffic at **Open Systems Interconnection** (**OSI**) layer 4 and the Application Gateway service operates at layer 7. This allows for routing of things such as **Transmission Control Protocol** (**TCP**) and **User Datagram Protocol** (**UDP**) traffic. If you want to filter on something such as an incoming URL, you will need an application gateway.

- **Azure Domain Name System** (**DNS**): The Azure DNS service provides a service for you to host your domain names in Azure for an annual fee. You can manage your DNS records with the same credentials you use to manage other services, all from within the portal or by using scripts.

- **Azure Content Delivery Network**: When you need to deliver software, firmware, updates, or other content, Azure **Content Delivery Network** (**CDN**) helps you do that by reducing load times and network bandwidth expenses. It is a global network for getting content to customers and employees.

- **Azure DDoS Protection**: Once your legacy workloads are running in Azure, you are going to need to protect them from cyberattacks and **Distributed Denial-of-Service (DDOS)** attacks. This is when a bad actor sets up an orchestrated attack that exhausts the application's resources so that it becomes unresponsive. Using real-time analytics and ML-based network traffic profiling, Azure can help detect and shut down these attacks so that your application experiences no downtime.

- **Azure Traffic Manager**: Azure Traffic Manager is a DNS-based traffic load balancer. If you have multiple deployments for the same application, Azure Traffic Manager allows you to distribute the application traffic how you want across those deployments. With this tool, you can optimize public network traffic to your applications.

- **Azure Firewall**: If you have to manage or work with an on-premises network environment, you no doubt have had to use a firewall to properly secure your environment. Azure Firewall is a cloud-native firewall built to secure and protect Azure workloads. It is an intelligent network firewall security service that is stateful and highly available. It is good to use in situations where you need to traffic to your migrated workload over the public internet. It comes in three configurations:

  - **Azure Firewall Standard**

  - **Azure Firewall Premium**

  - **Azure Firewall Basic**

  You will need to look at the features and decide which is the one for you. It also has a service called **Azure Firewall Manager** that allows you to manage your firewalls across your organization.

- **Azure Bastion**: If you decide to deploy your legacy workloads to Azure using VMs, at some point, you will need to log in to them and manage them. There are a number of ways to do that, whether you are using Linux or Windows using **Secure Shell (SSH)** or **Remote Desktop Protocol (RDP)** respectively. To simplify this, Microsoft created Azure Bastion. It allows you to connect to your VMs, be they Linux or Windows, using your preferred browser and the Azure portal, or native SSH or RDP. If you need to log in to various VMs to check or manage VMs, you will find this a useful tool that allows you to manage them from pretty much anywhere, provided you have a browser and access to your Azure portal.

- **Azure Private Link**: Azure Private Link is very useful and provides an extra layer of security when you need to get access to a private endpoint such as a service or storage endpoint. With Azure Private Link, the traffic between your VNet and the service endpoint travels over the Azure backbone network and not the public internet. This is great when you need to expose endpoints from your application to users or customers.

So, we have looked at the services that are used to connect and protect your network within Azure. Now, let's look at the services required to connect your Azure migration workloads to the outside.

Looking at Azure networking services for connection to on-premises data centers

These are the Azure networking services for connection to on-premises (or in-space) data centers:

- **Azure ExpressRoute**: In the world of legacy migration to Azure, probably the second most important network service you will need will be Azure ExpressRoute. There is a lot that could be written about ExpressRoute since it is rather complex, but here, I will provide a very concise overview.

  Azure ExpressRoute allows you to extend your on-premises private data center to the Azure cloud. In a nutshell, that is what it does. You will need it for your legacy transformation journey. It provides a private connection to an Azure data center and is a fast and secure way to connect to migrated legacy workloads. If you do not leverage ExpressRoute, the traffic ingress and egress for your application workloads will go over the public internet, which is not optimal. One thing to point out here is that what you are paying for when you buy Azure ExpressRoute via the Microsoft Azure portal is the provisioning of a connection endpoint to an Azure data center—that's it. There is another (and, typically, more expensive) component and transaction that needs to take place with your **telecommunications company** (**telco**,) such as AT&T NetBond or Equinix. The good news here is that if your organization is already using Azure in some capacity, then you might already have ExpressRoute set up, which will make your transformation journey simpler and faster.

- **Azure VPN Gateway**: If you decide that Azure ExpressRoute is too expensive or you do not want it, the next best thing is Azure VPN Gateway. It provides a lot of the same functionality as Azure ExpressRoute, but the traffic does go over the public internet. This might be okay since it is tunneled via TCP/IP, but if your organization has high security standards, this is something you will need to consider.

- **Azure Virtual WAN**: When you find that you need to support multiple VNets, ExpressRoute, and perhaps an Azure VPN Gateway all within a hub-spoke topology, you might want to look at the Azure Virtual WAN service. It allows users to seamlessly connect to various Azure resources using a mesh that is implemented in standard Azure connection points.

- **Azure Orbital Ground Station**: If you have the need to support users or applications that are in outer space, you can leverage Azure Orbital Ground Station. It allows communication between your Azure estate and spacecraft and satellites.

So, as you can see, there are a number of Azure networking services to address just about every network design pattern you need. Primarily for legacy migration purposes, you will focus on Azure Virtual Network and ExpressRoute. However, when you need to control access to these workloads or you need to integrate with other systems within Azure, you find the other services very useful.

Now, let's take a look at another fundamental area—Azure database offerings.

Azure databases

When it comes to database offerings in Azure, there are essentially two types you need to consider:

- IaaS database offerings
- **Database as a Service (DBaaS)**

IaaS databases and standard database servers that you are probably familiar with are ones such as SQL Server, Db2, and Oracle. These all run just as you would expect in Azure VMs in your Azure estate. To install them, you need to provision a VM with the proper OS configuration and install the database of your choice. If you have a substantial investment in Db2 and as a result have employees who are very familiar with Db2 **Database Administer (DBA)** administration and management, you can migrate your Db2 in z/OS to Db2 in Azure, which is Db2 LUW (which stands for Linux, Unix, and Windows). It should be noted here that Db2 for z/OS and DB2 LUW are different products and have different performance characteristics. The point here is that you can deploy them in Azure just as if you were deploying them on-premises.

There are some things to consider here that are actually non-technical. It has to do with licensing. You will need to look into the licensing caveats of moving your database (especially IBM databases such as Db2) to Azure since sometimes IBM provides a discount for running its software on its hardware and that discount could possibly go away when you migrate.

DBaaS database offerings are the new up-and-coming database offerings that provide some substantial benefits over their IaaS counterparts. These benefits are in the realm of automated management and performance. The underlying performance is typically not that much different since the underlying database is essentially an implementation of the IaaS offering; it just has more tooling to make it easier to operate, manage, and administer. In my conversations with customers who are migrating workloads to Azure, one of the more important aspects that DBaaS offers is that of staffing resources necessary to administer the database. Since DBaaS offerings have the technology to automate this, many times it takes fewer people to operate and manage the implementation, which can be a big cost saving for the customer. Let's take a look at some of the more popular offerings:

- **Azure SQL Managed Instance (MI)**: Azure SQL MI provides the most complete functionality of the IaaS SQL Server with a level of automation that reduces the management overhead usually required for a SQL Server implementation. It comes in two service tiers, as follows:

  - **General Purpose**: Provides typical performance for most business applications with built-in HA.

  - **Business Critical**: This offering is for applications that require high performance, high I/O, and the most resiliency. Depending on the nature of your migrated legacy workload, if it is mission-critical, you will probably want to select Business Critical.

- **Azure Database for PostgreSQL**: One of the trends I have personally seen in the last 2 years has been an increase in the number of customers who want to migrate their legacy workloads to PostgreSQL, especially for Oracle databases. I was curious, and I looked into it. It turns out that even though PostgreSQL does not have the full complement of feature functionality offered in a database platform such as Azure SQL MI, it has enough enterprise-class capabilities to be a viable database for legacy workloads. Keep in mind one of the things that goes along with this is the fact that for some of those advanced features that you find in Azure SQL MI—say, log shipping—you need to rely on open source solutions, but if you are okay with that, it provides a fast and highly stable database platform for enterprise workloads. Azure Database for PostgreSQL is a fully managed offering on open source database that is very cost-effective with predictable performance.

Other DBaaS offerings are available that deserve mention and that you should be aware of. They are listed here:

- **Azure Cosmos DB**

- **Azure Database for MariaDB**

- **Azure Database for MySQL**

- **Azure SQL Database**

Now that we have looked into the database options for your migrated workloads, let's look at one of the Azure services that bind them all together with Azure security and identity.

Azure security and identity

Consider that up to now in this chapter, we have talked about various Azure services that provide essential functionality for your legacy migration journey. Arguably, one of the most important services that keep these resources secure and provide access, or deny access, to these Azure resources is **Azure Active Directory** (**AAD**). It should be noted that Microsoft AAD is now called **Microsoft Entra ID**. The first thing you need to know about AAD is that it is a completely different service than **Active Directory** (**AD**), which was the on-premises directory service that was the center of Microsoft on-premises-based implementations in the days preceding Azure and until this day. Active Directory was designed for on-premises data centers and used Kerberos and **New Technology Lan Manager** (**NTLM**) for its authentication implementation. AAD was designed for cloud-based environments and used **Open Authorization 2** (**OAuth2**), **Security Assertion Markup Language** (**SAML**), and **Web Services Security** (**WS-Security**) for its authentication.

There have been many books published on AAD since it first came out in 2013, and the scope of the topic is way beyond the intention of this book. Instead, what you need to know is that everything—and I do mean everything—in Azure relies on AAD for access and creation of resources. It not only allows for access to cloud-based resources but it can also be connected to your on-premises Active Directory implementation for access to private network resources as well. This includes your mainframe and legacy estate as well. This is done with additional tooling, but it allows integration with identity systems such as IBM's **Resource Access Control Facility** (**RACF**) and Top Secret. It is very flexible with provisions such as AAD extensions, which allow AAD to provide services for migrated mainframe applications. It is also fault tolerant and easy to manage via the Azure portal and/or scripts.

Understanding the Azure approach to hybrid

As we mentioned earlier in the chapter, when Azure was first released, it was very focused on cloud and web development, with really no provision for running any other types of workloads. That all changed, however, in 2013, with the advent of ARM. Since then, Microsoft has put a lot of investment into hybrid options, with a focus on technologies that allow seamless management across on-premises deployments, Azure, and other public clouds.

Typically, when we refer to hybrid environments, we are talking about one of the following three scenarios:

- **On-premises environments**: Here, the scenario is to extend the on-premises data center into the Azure cloud. Technologies that you can leverage for this include the following:

  - **AAD**: To integrate with your on-premises Active Directory.

  - **Azure Monitor**: This allows you to monitor not only your Azure resources but also your on-premises resources as well. Note there is an agent that needs to be installed.

  - **Azure Arc**: This allows you to manage your on-premises resources from one pane of glass within Azure. It actually allows you to manage resources in other public clouds such as **Amazon Web Services** (**AWS**) and **Google Cloud Platform** (**GCP**) as well.

- **Edge environments**: Probably not within the scope of this book, but something to be aware of, is that there are provisions within Azure to monitor and manage edge deployments to leverage AI and ML for IoT and other types of workloads.

- **Multicloud environments**: In my current role supporting legacy migrations, I am seeing more and more interest in multicloud deployments. To be clear, what we are referring to here is customers who have opted to deploy some of their workloads in different public clouds. A good example would be customers who have workloads in AWS and Azure and GCP. As we mentioned before, you can manage and monitor these workloads with Azure Monitor and Azure Arc respectively. This is important for a couple of reasons, but probably the most obvious reason is that it provides a level of resilience.

So, the Microsoft approach to hybrid is really to extend the existing on-premises private data center into the Azure cloud, leveraging the additional functionality and value that the cloud provides and still being able to integrate with the resource in the private data center.

So, now that we have looked at how Azure secures resources in the cloud and on the edge of the cloud, let us look at how we deploy and maintain systems in the Azure cloud platform.

Deploying and maintaining systems on Azure

There are a number of ways to deploy and maintain applications in Azure. The most important part to take away from the section is that everything can be deployed in Azure via scripting. This scripting can be in the form of the following:

- ARM
- JSON templates
- Using a script extension for PowerShell
- CLI/Bash scripting
- Azure landing zone accelerators for mainframe workloads

You can also use **Terraform** and **Bicep** for the creation and deployment of Azure resources.

Another way, and probably the easiest, is to use the Azure portal, but keep in mind not every deployment feature is implemented through the portal. The portal is configurable and customizable. One of the features that I like about the portal is that when you are using it to deploy Azure resources, once you have the configuration that you want, you can transform that into a JSON template for future use.

Here is a screenshot of how you would deploy an application in Azure using the Azure portal:

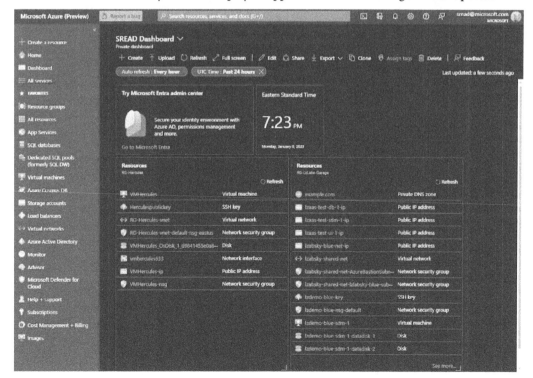

Figure 5.4 – Azure portal

You might have noticed that everything we have covered so far has an on-premises equivalent. What about all of the *cloud-native* stuff? Let's look at some of the more important cloud-native services you are likely to use.

Looking at cloud native, serverless computing, and microservices

There are quite a few terms being used these days in regard to public cloud platforms. I would like to address three that are very prominent right now: *cloud native*, *serverless*, and *microservices*. You will hear these terms pop up in various conversations referring to architectures and methodologies. I'd like to provide some clarification on these since they are quite often conflated with other terms and it helps to know what someone is actually meaning when they use these terms. Let's explore them now:

- **Cloud native**: Cloud native is a somewhat generic term used to describe an architecture that leverages cloud-based services. The distinction to be drawn here is that IaaS-based services such as VMs are not used or used very little. The benefit for the developers and IT staff is that it gets them out of the care and feeding of things such as the operating system and foundation services such as a database or message broker. Cloud-native applications use PaaS and serverless components to build the solution.

- **Serverless**: For serverless computing, customers and developers are only paying for the services they use. This gets developers and architecture out of the business of setting up things such as application servers and database servers when they really only need a very fast and reliable way to read and write messages. This is obviously more cost-effective, and all of the services that are offered are scalable and durable. You might think that there is really no server behind the scenes for something called "serverless." This is not true. The backend offering resides on a VM someplace in Azure, but it makes no difference to the end consumer of the service since they are shielded from the administration and management of it.

- **Microservices**: A microservice is a type of cloud-native architecture where what we would call an application is actually a collection of well-orchestrated services, each with a specific well-defined functionality. Microservices are developed and deployed as containers independently of one another. These services are loosely coupled and have their own technology stack with data and data functionality. This is a very different approach from creating applications that were used on legacy mainframes, where every resource was on the same hardware right next to each other and performance between the resources was extremely fast. This is what we will explore in *Chapter 7*.

So, there you have it. Hopefully, this explains these terms in the context of the cloud model we defined at the beginning of the chapter. You will no doubt hear them if you are embarking on a cloud modernization journey.

Summary

So, as you can see from the brief overview of the Azure platform, there are many tools available for your migrated workloads that you can leverage to make them high performing, scalable, and secure. In this chapter, we focused on the following topics:

- A brief history of Azure
- Various deployments of Azure, both locally and worldwide
- Azure compute, networking, and storage
- Azure database offerings
- Azure security and identity
- The Azure approach to hybrid cloud computing
- How to deploy applications on Azure
- Azure cloud-native services

In *Chapter 6*, we will dive deeper into how to build **fault tolerance** (**FT**) and **business continuity** (**BC**) into your Azure workloads.

6

Azure Cloud Architecture Overview for Mission-Critical Workloads

In *Chapter 5*, we looked at the Azure platform as a whole. In this chapter, we are going to focus on the features and services that support mission-critical workloads. Typically, these are the types of workloads that run on legacy systems, especially mainframe systems.

This chapter covers the following topics:

- What is mission-critical?
- Understanding Azure at a high level
- Understanding Azure's approach to availability and resiliency
- Looking at Azure's approach to **IaaS**
- Discussing firewalls versus network security groups
- Using Azure **PaaS** features
- Understanding Azure features for integration
- Understanding Azure features for backup and recovery
- Looking at a sample Azure deployment for a mainframe workload

What is mission-critical?

We will start with some semantics here, starting with what it means to be mission-critical and the ways to achieve this in modern **Information Technology (IT)** architectures. If you look up the definition of *mission-critical*, you will find that it means something vital to the function of an organization. In the case of software systems, we can probably all agree on what those types of systems are and why they need to be running as close as possible to 100% most of the time. If we peel the onion on this a little bit, we will discover that there are different levels of mission-critical. They are typically defined by two key metrics:

- **Recovery Time Objective (RTO)**: Simply stated, the RTO is the amount of time that the mission-critical system can be allowed to be out of service. This implies that after a certain point in time, there will be severe business implications if the system is not performing. As we will see, there are different approaches for different results.

- **Recovery Point Objective (RPO)**: This term is related to the amount of data that can be lost. Imagine a mission-critical system where you have a database that is processing transactions and then incurs a downtime event. RPO is the number of transactions that you are willing to not process essentially. They will be gone. For a retail store, we want this to be at zero, or as close as possible to zero.

So, these two metrics let us dial in our level or mission-critical, so to speak. Based on our level, we can decide the technology approach we want to use. And, you guessed it, the more we require 100% uptime, the higher the cost and effort. This would mean a very low RTO and RPO. Before we get into how we achieve this in Azure, let's get some more definitions out of the way for clarity. There are some very important approaches that you will hear being used when you talk about mission-critical workloads and RTO and RPO. They are fully redundant, fault-tolerant, **business continuity, and disaster recovery (BCDR)**. Let's take a look at them in some more detail:

- **Fully redundant**: You can think of a **fully redundant** system as the high bar of a mission-critical system. It implies everything is redundant and also on a redundant power supply. Let that sink in a little. When you think about it, the cost to implement a system like this is very expensive. You can almost achieve this in Microsoft Azure, but at the time of writing this book, Azure only supports one network fabric. Having said that, this can be achieved using the Azure Skytap implementation of VMware, which does support two network topologies.

- **Fault-tolerant**: A fault-tolerant system is achievable in Azure. It is a little less restrictive than a fully redundant architecture. It is an architecture that, as the name implies, can tolerate a fault or component outage without losing data (or very minimal data) and with no real implication to the user. An aspect of a fault-tolerant system is its resiliency, which addresses how many faults the system can tolerate before being inoperable. We will look at how to achieve this later in the chapter.

- **Disaster recovery**: Disaster recovery is a more general term to describe the ability to recover from some type of disaster either natural or man-made. It runs the spectrum from fully redundant to good old-fashioned tape backup. With fully redundant, you guarantee that no data is lost. With tape backup, you can lose some data.

- **Business continuity**: The term business continuity is even more inclusive and addresses how the systems that are running are designed. It also includes the staff and personnel who run the systems. For the remainder of this chapter, we will focus on the technological aspects within Azure that allow us to achieve a reliable mission-critical system.

Now that we have discussed the semantics, let's take a step back and look at Azure at a high level.

Understanding Azure at a high level

To achieve a well-architected mission-critical system, the best way to think about the Azure platform is as a collection of computing, storage, and networking components that can be consumed in several different ways, such as IaaS, PaaS, and **Software as a Service (SaaS)**. So, you can approach creating a mission-critical architecture just as you would if you were creating a highly available system on-premises, with only a few exceptions primarily related to having a public cloud versus a private cloud.

You might have heard the expression from Microsoft that Azure is *the world's computer*, and in many ways, this is true since it is composed of foundational services grouped into compute, networking, and storage. So, to make a system able to support mission-critical workloads, we have to make sure we put the right components together and leverage the native services within Azure that allow this. A key point to understand is that Azure was designed from the beginning to support high availability and mission-critical workloads. The more you leverage *cloud-native* Azure services, you will find that the necessary technology and features have already been included in that service. But the more you do in Azure that is IaaS-based, you will need to know what features need can be leveraged to achieve that same level of resiliency.

Just as a computer system has the basic compute, networking, and storage, it also has to have a way to provide resiliency and business continuity. Let's look at these features in more depth.

Understanding Azure's approach to availability and resiliency

As we mentioned previously, availability and resiliency were designed into Azure from the very beginning. A good example of this is how storage is presented to the user. If you create a managed disk in Azure that you want to use to store data, it will appear as one disk that you can mount and use. Azure maintains, at any given time, three copies of that disk for built-in, behind-the-scenes redundancy. The same is true for other storage offerings as well.

This is a good example in that it shows that the fundamental services that you rely on within Azure have availability and resiliency built in from their inception. But as we move up the application stack of services, we will require more features to provide the same level of coverage. Let's start with a simple example that you have probably encountered before. If I deploy an application to an Azure VM, that's great, but if that VM goes down, so does my application. If I want more resiliency, I need to choose different deployment models. That is what we will talk about in this chapter.

Looking at redundancy in Azure

When you are considering the level of redundancy needed to ensure availability in the event of an outage, you need to consider the scope of the outage. Do you just want to support outages that are local to the Azure data center, the Azure region, or the geographic location? Your level of support will help you decide which technology and features you can use. But before we get to that, it's time for a quick review of *Chapter 5* regarding Azure regions and data centers. Remember, an Azure zone is composed of one or more data centers. Each on independent power, cooling, and networking. An Azure region is composed of at least three Azure zones. You will also hear the term Azure location to refer to an Azure region. After the Azure region, we are in the realm of separate geographies for the Azure region. A good example of this is *Azure East* versus *Azure West*. The Azure regions are networked together by Azure's high-speed networking backbone so that you can fail over from one region to another. Now, let's take a look at how we can dial in the level of business continuity we want at the Azure region level.

Looking briefly at Azure SLAs and the nines of availability

Microsoft has different **service-level agreements** (**SLAs**) for many of its Azure services. I highly recommend that you check out the following link for more information: `https://azure.microsoft.com/en-us/support/legal/sla/summary/`. What this means is that Microsoft is willing to guarantee uptime for a given resource and financially back it up – excluding indemnity, which means that they will back the cost of the resource for the downtime, but not the business implication of the downtime. Rather than get into a lengthy discussion about the various SLAs for various **Stock Keeping Units** (**SKUs**), we will focus on the intent of this SLA, which is to have a guarantee with the user of Azure that the service they are using has been developed and engineered to the necessary standards to support that given level of availability. In Azure, it is possible to have your workloads supported to four and five nines of availability. To realize what this means, let's look at what this equates to in time per year for an outage, as shown in the following table:

Amount of Nines	Percentage Uptime	Downtime Per Year
Two nines	99%	3 days and 15 hours
Three nines	99.9%	8 hours and 45 minutes
Four nines	99.99%	52 minutes and 33 seconds
Five nines	99.999%	5 minutes and 15 seconds

Table 6.1 – Nines of availability in uptime

As you can see, having a system with this level of availability is impressive and hard to achieve on an on-premises private data center without significant investment.

Looking at redundancy within a region

Typically, the legacy and mainframe workloads that you are migrating to Azure, will execute in either a **virtual machine** (**VM**), a container in **Azure Kubernetes Service** (**AKS**), or Azure Container Apps. This is not to say that they cannot be deployed into other services such as Azure Service Fabric, they can. But, as we have said before, if you deploy to cloud-native services, resilience is typically built into that service. So, let's focus on what we need for VMs, AKS, and Container Apps.

Within an Azure region, you can have highly available workloads – you just need to make sure you are leveraging the proper features and services. Let's take a look at what they are:

- **Azure availability sets**: An Azure availability set is a logical arrangement of VMs that allows Azure to manage them appropriately to provide redundancy and availability. You can achieve 99.95% uptime via an Azure SLA if you deploy a VM using availability sets alone. To meet this SLA, you need to deploy two or more VMs within the availability set. There is no cost for the availability set itself, only for the VM instances. Azure assigns an update domain and fault domain for every VM in the availability set. You can think of a *fault domain* as a separate server rack on a separate power supply and network switch. What this does is it spreads the likelihood of failure over multiple server racks to provide redundancy and resiliency from power failures. An *update domain* provides similar functionality for updating the VM. From time to time, patches and updates will need to be provided for the VM and you will want to do this without incurring downtime. Update domains provide a way to not incur the downtime typically required.

- **Azure availability zones**: An availability set takes place within an Azure data center. An entire Azure data center can have an outage. However, it is very unlikely. It is also possible for an entire Azure region to incur an outage, though that is even more unlikely. To defend against a data center outage, you will want to leverage Azure availability zones. They expand the level of redundancy across Azure zones. Remember that a zone is composed of no less than three Azure data centers so now, we are at the region level. Each Azure zone has a separate power source, cooling, and networking. What this means is that if a given data center goes out, the workloads will keep running in the other Availability Zones.

- **Virtual machine scale sets**: VM scale sets provide both scalability and availability. With this approach, you can create a group of load-balanced VMs. The VMs need to be the same type and this architecture implies that they are running the same workload. The number of VMs can grow or shrink based on the parameters you define and they can all be tied to a schedule. So, if you want to scale to 100 VMs around Christmas, but all other times of the year you do not want to scale past 60, you can set this up in your deployment configuration and Azure will enforce the scalability rules for you. Typically, you would leverage this for workloads that are dynamic and elastic.

- **Azure Load Balancer**: As the name implies, Azure Load Balancer allows you to balance incoming loads between two or more VMs, be they deployed in an availability set, availability zone, or a scale set. It operates at level 4 of the OSI and is the endpoint for the client. It can distribute the load based on predefined rules, such as VM capacity or round-robin. The load balancer is a key component in creating a highly available cluster using IaaS.

- **Azure storage redundancy**: We looked at this earlier when we talked about how Azure keeps three shadow copies for you to improve resilience and protect against hardware failures. Let's take a deeper look right now. The baseline for this level of redundancy is the three copies we talked about, but some offerings let you improve on this:

  - **Locally redundant storage (LRS)**: LRS keeps three copies of your data within a single data center in an Azure region. It copies your data synchronously to the different copies. It is the least expensive but is not recommended for workloads that require durability and high availability.

  - **Zone-redundant storage (ZRS)**: Much like Azure availability zones operate at the zone level, so does zone-redundant storage. It makes copies of your data to the three separate zones within the data center. By doing this, it provides 11 nines of durability for your data within the region.

  - **Geo-redundant storage (GRS)**: GRS takes this copying of the data to the next level and spans it geographically. It leverages LRS to keep the data copied and in sync at the primary Azure region, but then copies the data to a secondary region where the LRS for that region takes over and does the same copy process. This is similar to how Azure availability zones operate at the zone level.

  - **Geo-zone-redundant storage (GZRS)**: The only difference between GRS and GZRS is that GZRS uses ZRS locally in the Azure region to copy the data, so it is technically more resilient across Azure regions.

Now that you have a basic understanding of the different constructs that you can use to get your level of resiliency, let's see how easy it is to leverage these features. I want to do this because I want you to understand how easy it is to use them and the consequences if you do not. Consider this – if you go to the Azure portal and create a VM and do not leverage the features we have talked about, your VM is really no different from a PC you might have running in your office or a desktop. If you incur an outage or somebody turns it off, the workload goes down. When you are running a mission-critical application, this is not an option.

The following screenshot shows how to create a VM in Azure:

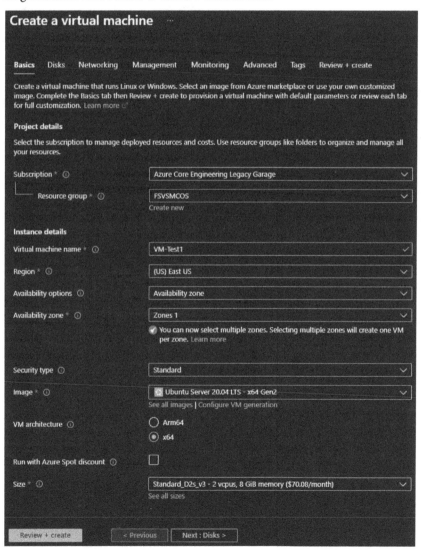

Figure 6.1 – Creating an Azure VM

Notice the section devoted to **Availability options** and **Availability zone**. This is where you will need to select the features we just talked about for the VM to be highly available within the region. It's that simple. There is another section devoted to setting the type of storage you can select. From a cost standpoint, there are a couple of points to keep in mind when you are selecting storage options within an Azure region. Data transfer within an Azure availability zone is currently free of charge; data transfer between availability zones is not.

Keep in mind that everything we are looking at here can also be implemented via the **command-line interface** (**CLI**) and using your favorite IaaS scripting languages, such as ARM templates, Terraform, or Bicep.

So far, we have focused on resiliency within a region. Now, let's look at how we can make our application resilient across Azure regions.

Looking at redundancy across regions

Now that we have looked at redundancy and resiliency within an Azure region, let's look at how you can create a system that leverages multiple Azure regions to attain BCDR. This architecture protects against the unlikely event that an entire Azure region has an outage. With this scenario, we need to make sure that we have at least two copies of the mission-critical workload in each region and we need to ensure that both systems are kept in sync. To do this, we need to use **Azure Site Recovery** (**ASR**).

ASR is the service offering you will need to use when you need a full-featured BCDR implementation in Azure. It replicates VMs across Azure regions and can even be used for replication with on-premises servers as well. There are a few things to keep in mind when leveraging ASR. The first is that you can and will experience some downtime. This is important to realize. Unlike fault-tolerant, highly available systems that are redundant and can experience a fault and keep on running without the user experiencing any downtime, the value of ASR is that even though there will be a period, usually very short, where the system will be unavailable to the user, there will not be any loss of data and the system will be up and running in a secondary data center in a short period. This period varies based on the complexity of the system. For an application with 8 to 10 VMs, this will usually take anywhere from 5 to 20 minutes, depending on things such as the network's speed and the size of the VMs.

To do this magic trick, ASR keeps all of the VMs that are a part of the application in sync with each other. This means that any changes to the operating system or any other relevant system on the VM are replicated to the enlisted VM. As I mentioned, this can be done for VMs in different Azure regions and also for VMs that are on-premises in your private data center. This is a good example of hybrid cloud computing.

The other crucial task that ASR performs is monitoring the health of two systems. In an ASR deployment, one system (production) is active and the secondary BCDR system is passive, ready to be lit up for transaction load when necessary. There is essentially a heartbeat between the two systems. Once ASR detects that the primary production system is down, it starts the recovery process on the secondary site. The process typically follows this pattern, but it can be customized:

1. The process initiates the database for incoming transactions. For a SQL Server AlwaysOn or Azure SQL MI implementation, this means allowing the database to accept incoming transactions. While in passive mode (that is, production is up), the database is running, but all of the transactions that it receives are coming from the production system. It can be used for reporting, but it cannot accept transactions on its own.

2. The application tier servers are started up so that they can also accept incoming transitions. It also does this for the other tiers of the application.

3. The main **Domain Name System (DNS)** entry is set so that the new up-and-running system is open for business. At this point, all inbound traffic for the application can be processed without any loss of transactions.

The following figure shows this:

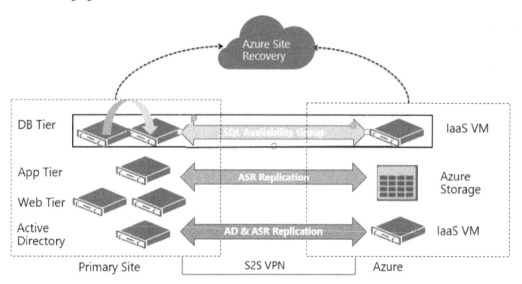

Figure 6.2 – Azure Site Recovery

There are a few things to point out that might not be very obvious with this solution. First, ASR is *not* used for the data tier. It does not guarantee the transactional consistency of the database. You will need to leverage the proper technology within the database that you choose to attain this. If you decide to use SQL Server MI Business Critical, this is a feature that is inherent in the offering. If you decide to use Oracle or IBM DB2, there is an equivalent technology that achieves pretty much the same thing. IBM leverages a solution called HARD within DB2 and Oracle has a suite of technologies within the Golden Gate solution. Also, typically for the presentation tier, you should probably leverage Azure App Service. This is a cloud-native service that's perfect for hosting presentation-tier workloads that have built resilience across Azure regions.

With that, we have looked at how we can provide availability and resiliency both within and across Azure regions. Now, let's look at the features we can use for IaaS.

Looking at Azure's approach to IaaS

When you are migrating legacy applications to Azure, IaaS is often considered the first step. This is usually because migrating the existing workloads involves the least amount of risk, but not always. Whether you are rehosting or refactoring, IaaS can offer some benefits when it comes to the first step in your modernization journey.

For some organizations that are modernizing application workloads, the steps of the journey look like this:

1. Rehost or refactor to Azure VMs.

2. Modernize to cloud-native services where it makes sense.

This is often a good approach for modernization because it allows you to leverage the other benefits of Azure, such as its analytics platform, and also lets you say goodbye to your expensive legacy hardware.

Looking at IaaS for data

With only a few exceptions, you can deploy any x86-based database into an Azure VM. Good examples of this are Microsoft SQL Server, IBM DB2, and Oracle DB, as well as a host of others. Each of the database solutions that you decide to leverage will run just as it did on-premises. Each of the database solutions also has a technology for resiliency that you will need to leverage for transaction consistency in the event of a failure.

So, for IaaS databases in Azure, this will be very similar to what you have already experienced with your on-premises installations. Now, let's look at the applications in IaaS.

Looking at IaaS for applications

We have covered many of the offerings related to IaaS for applications, mainly VMs. But what happens when we need to scale that VM either up (that is, into a larger VM) or out (that is, by adding additional VMs to handle the load)? Let's address the first scenario. Either in the Azure portal or via scripting, you can change the size of an Azure VM. Keep in mind that the VM will need to cycle (that is, reboot), but if this is acceptable, it is very easy to do. You can even monitor the load on a VM and do this automatically. But many times, you will not want to incur a small period of downtime. This is where VM scale sets come in. We briefly introduced these in *Chapter 5*, but now, let's look at them in a little more depth.

VM scale sets allow you to dynamically add capacity to a load-balanced group of VMs. These VMs are running the same application and are similar server offerings. The nice thing about scale sets is that no downtime is necessary and you can add the capacity up to a user-defined limit. When the load has passed, you can scale in much the same way. This offers availability and scalability for IaaS applications. If you want functionality with a VM implementation that is similar to some of the benefits of Kubernetes, you should evaluate VM scale sets. It should be noted that the application needs to be developed to support this ability. This typically means that they need to be stateless and not have server affinity (that is, a process that ties to a particular named instance of a VM).

Discussing firewalls versus network security groups

In my travels working with customers and creating Azure deployments, there is one topic that often comes up concerning Azure networking. Before we get into the debate, however, let's define some terminology. The most important artifact in the Azure networking tool bag is *Azure VNet*. We defined them in the previous chapter, but it is worth mentioning them here for this discussion. Consider that you cannot create an Azure VM or any other Azure artifact that requires networking unless it exists in VNet. This is because the VNet is the network container that allows you to group Azure resources. You can subdivide a VNet into subnets and further group resources within them as necessary. An Azure **Network Security Group** (**NSG**) allows you to filter traffic between these resources using security rules. This sounds a lot like a firewall, doesn't it? Azure offers *Azure Firewall Service* and third-party firewall solutions too. Without getting into the details of the debate, the consensus is that while firewalls offer some advantages over an NSG in some scenarios, for the most part, you can safely lock down your deployment by leveraging Azure NSG. However, keep in mind that while an NSG does basic filtering like a firewall, it does not filter layers 3, 4, and 7 traffic. NSGs also do not use threat intelligence filters, which are typically used for external incoming traffic.

Using Azure PaaS features

We described IaaS, PaaS, and SaaS in *Chapter 5*. One of the big advantages of PaaS offerings in Azure is that much of the resiliency both in a region and between regions is already built in and all you need to do is leverage them. Let's look at some of the benefits of using PaaS offerings in Azure:

- **Faster implementation time**: Since the PaaS offering is already set up, typically, all you need to do is select the service and configure it.

- **Reduction in complexity**: Normally, if you are configuring a similar IaaS implementation of a PaaS offering, it will be more complex, and you will be responsible for the underlying artifacts, such as the operating system and dependent services.

- **Improved quality of service**: Most PaaS offerings in Azure are backed by an SLA, which is a financially backed agreement that the service is reliable. Getting to that level of reliability takes both expertise and specialized infrastructure.

- **Increase in agility**: Most PaaS offerings offer increased agility by offering elasticity and being able to scale up or down to meet the load of the application.

So, as you can see, there are some pretty compelling reasons for opting for PaaS services when you are migrating to a legacy workload. Having said that, what is the trade-off? And some trade-offs come with selecting PaaS over IaaS. They typically come down to control and feature availability. We will point this out when we look at PaaS for data and PaaS for applications. If you think about it, this makes sense. When you are deploying your servers for either an application or a database, you have access to some features in that operating system or database that have been intentionally shielded from you in the PaaS offering. This is to bring you the benefits we just listed. The best way to decide when to use IaaS or PaaS is to evaluate the dependencies to make sure PaaS is providing everything you need. If it is not, then you will need to consider IaaS as an alternative.

A growing trend in cloud computing is the shift from IaaS deployments of databases to database PaaS. The reason for this is that database PaaS provide automated features that allow the database platform to respond to real-time events in an automated fashion, thereby reducing the need for database administrators. Indeed, in my experience working in modernization, I have encountered many customers who associate **full-time employees (FTEs)** with the IaaS deployment of a database. When you consider this, you will see that there are potentially substantial cost savings for an organization when they leverage database PaaS. Let's explore this in more depth.

Looking at database PaaS

In the world of relational and non-relational databases, in the case of open source offerings, the products that are on the market are very mature or becoming mature very fast. As we mentioned earlier, depending on the specific features and functionality you need, leveraging a **Database-as-a-Service (DBaaS)** offering over an IaaS implementation might make more sense. Let's go through an example where we'll look at an IaaS offering versus the corresponding DBaaS offering. To make this example straightforward, we'll look at SQL Server. At a high level, there are essentially two ways you can leverage SQL Server in Azure. First, you can set up your own SQL Server deployment in VMs. This is the IaaS approach. If you wanted to make that SQL Server deployment highly available and support business continuity, you would need to leverage *SQL Server AlwaysOn Availability Groups*. Not to get into the weeds with this but suffice it to say that you would need to set up at least three SQL servers in a specific configuration. This is not a trivial task and is best left to someone who has the expertise to do it. The other option is to use *Azure SQL MI (Managed Instance)* Business Critical, which is the DBaaS offering of SQL Server AlwaysOn Availability Groups. You can consider Azure SQL MI as a *native service* since it is offered to the user as a fully implemented, ready-to-use cloud service. So, hopefully, this example gives you an idea of why cloud-native offerings are becoming so popular and the value they bring to a cloud architecture.

You can deploy pretty much any database you like in Azure while leveraging IaaS, so long as it runs on either Windows or Linux and the x86 platform. So, databases such as IBM DB2, Oracle, and Adabas run just fine in Azure, just like they would in your on-premises data center. You will need to work out the licensing, of course, but the feature functionality of the on-premises and Azure offerings will be equivalent.

For the PaaS databases, the number of offerings is growing since there is a demand for the benefits they bring. PaaS database offerings are typically specialized for certain types of workloads, depending on what type of workload you are migrating. Keep in mind that each of the offerings we are going to look at can also be deployed as IaaS in a VM, just like you would do in your on-premises data center. Let's go through the offerings you will most likely need to evaluate. Keep in mind that there are other offerings as well, but they are very specialized:

- **Azure Cosmos DB**: If you have been in IT or development for the last 5 to 10 years, you have no doubt heard of NoSQL databases. These are different from relational databases in that they are unstructured. The proposition is that the application retains the ability to access the data and lessens the burden or the need for a relation engine. This type of database is in high demand, especially for cloud workloads. Azure Cosmos DB is the best of both worlds to a certain extent in that it offers both relational and NoSQL database access for modern applications. It is fully managed and, as you would expect, reliable and highly available.

- **Azure SQL Database**: Azure SQL Database is a PaaS database service in Azure that is fully managed and provides most of the critical tasks that you require from a database. It is built on top of SQL Server and is reliable and highly available. It is targeted toward web developers for web applications but can be used for other workloads as well.

- **Azure Database for MySQL**: You have probably no doubt heard of MySQL. It is a free and open source database platform that has gained popularity and has a growing population of developers that use it. Azure Database for MySQL is a fully managed database service that can handle enterprise workloads.

- **Azure Database for MariaDB**: Much like MySQL, Maria is another open source, free database that has become more popular recently. It can also be used in various enterprise-type workloads. Azure Database for MariaDB is a fully managed implementation.

- **Azure Database for PostgreSQL**: Yet another open source and free database platform is PostgreSQL. I have personally seen the demand for PostgreSQL growing among the customers I have worked with, especially for Oracle and DB2 workloads. It is reliable and highly available and with Azure Database for PostgreSQL, it is offered as a fully managed Azure service. Keep in mind, however, that Azure PostgreSQL does not support **User Defined Type** (**UDT**). So, if you are leveraging UDFs, you will need to use the IaaS deployment.

- **Azure SQL MI**: We went over Azure SQL MI earlier in this chapter. You can think of it as a PaaS offering of SQL Server with most of the highly reliable features you would expect from SQL Server.

There are some other PaaS service offerings in the database category that, while they are not exactly databases in the true sense of how you might think of a database, they support applications and databases for a specific purpose. They are as follows:

- **Azure Database Migration Service**: Allows you to migrate relational databases to Azure.

- **Azure Cache for Redis**: Not exactly a database, Azure Cache for Redis is an in-memory store based on Redis software. It can be used for extremely fast access to cached data. A scenario where this is used is in clustering when you need to make volatile data available to different nodes in the cluster.

Now that we have explored the database PaaS alternatives, let's look at the alternatives for PaaS for applications.

Looking at PaaS for applications

When it comes to PaaS for applications, there are even more offerings, so let's go over the ones that you will more than likely encounter. Keep in mind that these services are fully managed and can be selected from the Azure Marketplace:

- **App Service**: App Service is primarily aimed at developing web and mobile applications. For mainframe and legacy migration implementation, App Service can also be used for the RESTful API service facade that many mainframes already have installed in their on-premises data centers. It can also be used for any other type of user interface you need to support for the presentation tier. This is typically found in a VMWare implementation. App Service is fully managed and has auto-scaling and high availability.

- **Azure Kubernetes Service (AKS)**: Kubernetes has become very popular lately and it is a viable alternative to deploying legacy workloads in VMs. It allows a prescriptive environment for running containerized applications. AKS is, as you guessed it, a fully managed implementation of Kubernetes. It is resilient and can be deployed in a highly available configuration across Azure regions. As a fully managed service, it offloads some, but not all, of the complexity of running a Kubernetes application.

- **Azure Container Instances**: If you have ever worked with Kubernetes or even Azure Kubernetes Service, you know that it is not a trivial endeavor. You need to have the expertise and know what you are doing. Azure Container Instances is a response to this complexity and offers many of the benefits of AKS, in an easier-to-manage environment.

- **Azure Event Hubs**: For a legacy mission-critical system, typically, Event Hubs is, at least many times, the system of record. By this, I mean it is the system where the transactions take place that runs the business. Inevitably, we will want to send this data to an analytics engine for reporting and analysis. Azure Event Hubs was created for this purpose and allows you to stream data into your chosen analytics platform for analysis.

- **Azure API Management**: In the world of cloud and distributed computing, the ability for applications and services to be able to call other applications and services is the foundation. This is typically done using **application programming interfaces** (**APIs**). APIs abstract the implementation of an application or services to an agreed-upon contract for how they can be invoked and used by other processes. These APIs are typically part of an application gateway. As the name implies, Azure API Management allows you to not only manage and monitor these gateways but also quickly create and make them public in Azure, as well as externally to Azure.

Understanding Azure's features for integration

When it comes to integration, there are several services within Azure you can leverage, each with a particular focus, depending on what kind of integration scenario you are working on. Let's take a look at them:

- **Azure Logic Apps**: When you are creating applications that need to integrate with a variety of other cloud-based services or applications as well as on-premises applications, you will probably need to look at Azure Logic Apps. It allows you to create workflows so that you can integrate business processes. Remember **service-oriented architecture** (**SOA**) – this allows you to build those types of applications in Azure.

- **Azure Service Bus**: Another popular programming paradigm is **pub-sub** (**publish-subscribe**). This is where an application publishes messages to other applications. You might also need a service broker to ensure guaranteed delivery of messages. When you need that type of functionality, you need Azure Service Bus, which allows you to decouple applications to provide high availability and scalability.

Understanding Azure features for backup and recovery

Earlier in this chapter, we looked at ASR, which is the primary way to keep VMs in sync across regions. It can also be used within a region if needed. In addition to ASR, there is also Azure Backup Service. With *Azure Backup Service*, you can back up pretty much any storage you are using in Azure to LRS, GRS, or ZRS. You can back up Azure VMs, managed disks, Azure file shares, and Azure Blob Storage. You can also back up your databases. You can even backup storage in your on-premises data center, provided you have the *Azure Backup Server (MABS)* agent installed. The data is encrypted in transit and at rest, so it is secure.

Looking at a sample Azure deployment for a mainframe workload

So, based on all of the technologies and services in Azure that we have outlined, what would a typical legacy workload look like? The age-old answer to this question is, of course, it depends. But some consistent patterns allow us to put together a good starting point for a prescriptive architecture. The following figure shows what a mainframe workload would look like in Azure:

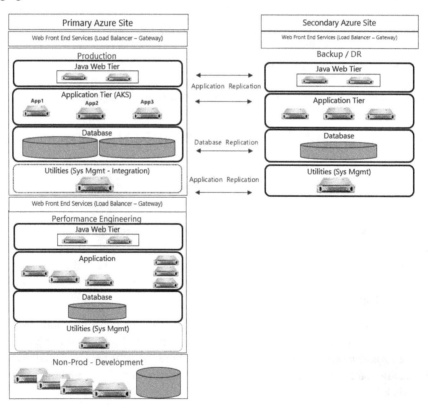

Figure 6.3 – An example of a legacy workload in Azure

These are often called **Azure landing zones**. There are a few things to point out about this architecture. First, as you can see, the production workload is made highly available within the Azure region by leveraging an application cluster that uses Redis Cache to keep the application servers in sync. This provides scalability and availability. With this type of implementation, more capacity can be added by adding an additional application server.

The database tier is made highly available in this deployment by leveraging SQL Server in an AlwaysOn cluster. This ensures that the databases both in the region and in the BCDR region are transactionally consistent. The VMs in the production environment are kept in sync with the VMs in the BCDR region via ASR. As we talked about earlier, keep in mind that the VMs in the BCDR region are passive until a failover occurs. Every box you can see in the preceding figure implies a networking boundary. For each tier, the best practice is to put each one in its own VNet and within those VNets, create subnets for the groups of VMs and/or services you use. Also, keep in mind that some Azure services such as Azure SQL MI require their own VNet to be deployed.

In many mainframe deployments, there is a performance engineering environment that is a replica of the production environment but without high availability. It is also similar to a UAT environment. For non-production environments, they are usually needed by developers who are writing new applications or maintaining existing applications.

It should also be mentioned that the production environment is assumed to be running 24/7 or 730 hours per month. What this means is that the services are never turned off. This affects the amount of Azure consumption. The performance engineering and non-production environments usually do not need to be on all the time. Typically, they are on for the workday and can be turned off when they're not being used. This can offer you substantial savings on your Azure spending.

Summary

The intent of *Chapters 5* and *6* was to provide a brief overview of the Azure platform and also show how you can use its features and services to put together a reliable and highly available deployment for a legacy workload. Hopefully, this has provided you with a better understanding of the Azure platform and how you can take your mission-critical workloads and deploy them in an Enterprise-class way.

The next step in understanding how to migrate legacy workloads into the Azure platform is to understand the differences between a monolithic architecture and a distributed architecture. This is what we will explore in *Chapter 7*.

7

Behold the Monolith – Pros and Cons of a Monolithic Architecture

We will look at monolithic systems in this chapter since most mainframe applications were developed for this type of environment. Currently, there are trends to move away from monolithic architecture in the cloud. However, the monolith approach served a purpose. We will look at whether a monolithic system should be redesigned for the cloud or whether you should host the monolith in Azure, which is also an option.

This chapter covers the following topics:

- What is a monolith?
- How did most mainframe applications become monoliths?
- Why are born-in-the-cloud applications not typically monoliths?
- What if you want to transform a monolith to cloud native?

What is a monolith?

For the purposes of this book, a monolithic application has the following characteristics:

- It is a single overall application where individual pieces have strong dependencies—in other words, a single overall code base where the code in modules does not function independently and changes to one module affect other modules.
- If required, it typically scales up rather than scales out.
- It depends on OS-specific services for functionality, making them tightly coupled to a specific vendor.

- It is difficult to separate functionality to modernize code a piece at a time, usually requiring a *big bang* or modernizing the entire mainframe at the same time approach to modernization. This includes testing required for the whole system.

- It is difficult to make changes in one place in the code without breaking functionality in a different place.

The obvious question that comes up deals with why anyone would develop a monolithic system in the first place. Here are some reasons why monolithic applications were created and still run today:

- Creating a monolith was the only option when the application was originally developed. Keep in mind that many of these systems were developed more than 10 and possibly 40+ years ago.

- Modern scale-up hardware can often meet the performance requirements for monolithic systems.

- Changes over time have made the monolith even more difficult to modernize due to increased interdependencies.

- If it ain't broke, don't fix it. Unless there is a compelling reason to modernize, many organizations simply avoid modernization and live with the technical debt of older systems.

- Many of the original developers are no longer available for organizations, making modernization even more problematic.

The remaining content of this chapter will look specifically at monolithic applications on IBM mainframes (z/OS) for strategies to modernize these systems to Azure. We will also look at IBM iSeries and Enterprise UNIX.

How did most mainframe applications become monoliths?

Monolithic systems are common on IBM mainframes for a number of reasons:

- Mainframes have been around for 50+ years and can still run applications developed 50+ years ago

- Common programming languages such as **Common Business Oriented Language** (COBOL) and **Programming Language/I** (PL/I) allowed for procedural code to be easily developed

- Mainframe tools such as **Customer Information Control System** (CICS) and **Information Management System** (IMS) allowed procedural code to be developed in an environment that could scale up easily

- Mainframe hardware has become more powerful over time to keep pace with the performance demands of monolithic applications

With these factors in mind, it is easy for monolithic systems to grow to over 10 million lines of interdependent source code, making the task of modernizing look very daunting and costly.

What about iSeries and Enterprise UNIX?

Monolithic systems are not unique to COBOL applications written for the mainframe. Monolithic applications were also developed for legacy OSs such as iSeries, Enterprise UNIX, Windows Server, and Unisys mainframes, even when using more modern languages such as Java and C#.

What are the monolith's advantages?

Monolithic applications do have some advantages. Listed ahead are some of them:

- Mainframe monoliths are typically self-contained; they are not dependent on other systems or services

- Monoliths, by their nature, typically scale up and are well suited to systems where faster processors can be deployed for increased scalability and performance

- Monoliths typically work well with transaction monitors such as CICS and IMS on mainframes, or Tuxedo on Enterprise UNIX

- Transaction models with monoliths are typically easier since they can easily support pessimistic locking and have less transaction contention

Monoliths were particularly suited to mainframe scale-up architecture that used **transaction processing (TP)** monitors. This type of architecture allowed developers to write procedural code in a self-contained environment and enabled programmers to write code that would be inefficient in a distributed environment. Additionally, when originally developed, the only option was to deploy an enterprise-class system on a monolithic-capable platform such as the IBM mainframe.

What are the monolith's disadvantages?

Monoliths also have their problems. Listed ahead are some of these:

- Being self-contained, mainframe monoliths often required developers to write duplicate code for each system. Languages such as COBOL do allow copybooks to share routines between applications, but each application has to compile the copybook within the program. There is no concept of shared services beyond what the TP monitors provide.

- Monolithic applications are difficult to scale out or use with a load balancer. Typically, if monoliths are scaled out on the mainframe, a **shared-everything approach** is needed, which requires a tightly coupled implementation such as the mainframe **coupling facility (CF)**.

- Monoliths are difficult to implement as shared services in a cloud environment or for leveraging existing shared services.

- Monoliths typically do not support on-demand scaling such as elastic compute. They depend on faster processors rather than adding additional processors or cores.

- Monoliths typically do not support opportunistic locking or compensating transactions and do not take advantage of containerized deployment options.

To summarize, monoliths do not usually take advantage of modern cloud environment features, other than features available in Azure for **Infrastructure as a Service (IaaS)** VMs, such as a large number of vCPUs and optimizations for memory or compute.

The following diagram highlights some of the aforementioned disadvantages:

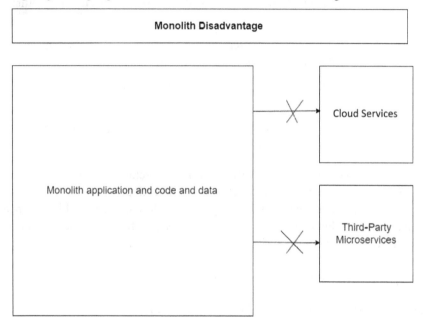

Figure 7.1 – Disadvantages of a monolith

Monoliths have both advantages and disadvantages. The key advantage is the *all-in-one* approach where there are limited dependencies on other applications or systems. The key disadvantage is the inability to leverage other cloud services and code. It's a trade-off on what is the right approach to take.

We will now look at the born-in-cloud approach.

Why are born-in-the-cloud applications not typically monoliths?

Azure can support both monolithic and cloud-native architectures. This section will look at both approaches, keeping in mind that cloud native is the preferred approach if possible.

Can you migrate a monolith to Azure?

The simple answer is *yes*. Azure provides VMs with the following capabilities that can handle monoliths in a scale-up architecture:

- Over 400 vCPUs for applications that can take advantage of multiple CPUs
- Compute-optimized VMs to provide fast clock speed or memory
- High **input/output operations per second (IOPS)** and data throughput for batch applications
- Proximity placement groups that allow minimal latency between VMs

With these capabilities, Azure can handle most mainframe workloads easily for both scalability and performance without taking advantage of Azure cloud-native features.

What if you want to transform a monolith to cloud native?

Now, we will look at creating a cloud-native approach for transformation from a monolith.

The tyranny of locality

Before we dive into the mechanics of what we need to consider when we are migrating from a monolithic architecture, let's consider the fundamentals. In a monolith, by definition, almost every resource we need to access is local, or very close to local. This means that the way you approach your development is different. In a **distributed environment**, this is not the case, hence the *distributed* part of the term. Typically, we need to negotiate a network and security handshake, which is time-consuming, and this is one of the essential differences between the two paradigms. Let's be very clear here: *no one model is better than the other, but they are fundamentally different*. In a monolithic architecture, we can get away with things such as polling (that is, querying a resource multiple times to detect a change of state) of resources, which leads to *chatty* applications. When we try this approach in a distributed environment, where the polling traffic has to travel over a network, the latency will soon let us know it is not an option in most cases. This is just one example; there are many others.

The database-centric nature of a monolith

Another fundamental difference between a monolithic architecture and a distributed architecture is the database. In the world of the monolith, the database is king, and there are usually only a few—or sometimes one. When you think about it, this made perfect sense. If I possibly have a large number of applications that potentially need to share data, let's put everything into a giant database where we can share the data. If I have separate applications, potentially for different business domains, then no problem; we can share the data through either the applications or procedures stored in the database. The processing to join the data takes place in the database and is extremely fast. This approach worked very well and was blazing fast.

However, where it started to break down was when, over time, growth forced the custodians of the application and database to upgrade and add new features or accommodate new business patterns. Why was this an issue? Because everything was linked together. The *Gordian knot* analogy comes to mind here, where you have created something so complex that very few people in the organization understand it. The other issue here was that everything was in one place. The database was the **crown jewels**—so to speak—of the organization. If some malcontent could gain access to the database, they could possibly have everything.

The following diagram depicts a monolithic architecture and the database-centric approach:

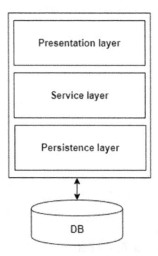

Figure 7.2 – Cloud-native data approach

Contrast this with a microservices approach, where every business domain has its own applications, processes, and databases. It is a radical departure that requires a different way of thinking about our applications. We will dive into this in more depth in just a bit, but for now, here is a diagram that shows the microservices approach so that you can contrast the two:

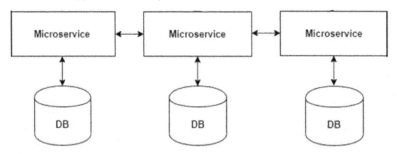

Figure 7.3 – Microservices with separate distinct databases

Let us now look into the *radical departure* we mentioned previously.

The paradigm shift

So, as we consider the pros and cons, let's consider what a monolith is good at doing. The answer is it is good at processing large amounts of records and transactions very efficiently. This inevitably led to processing patterns where the online systems recorded the transactions of the day and batch jobs processed those transactions at a later point into another system of record. This worked very well for many years. Contrast this with a distributed system approach where we could replace the need for batch processes with something such as **eventing**, where updates to the systems of record occur asynchronously as they happen. Hence, no need for batch processes. This is just one example where we can *reimagine* monolithic workloads in a more modern approach.

What makes an application cloud native?

So, why is application development for enterprise and web-based applications trending toward cloud native? And, more importantly, what makes an application cloud native? If you look up *cloud-native application* on your favorite search engine, you will see a definition similar to this: *An approach to building applications that leverages the advantages of distributed computing available in cloud platforms.* What this means is that now, for modern applications, we can really just focus on the application and not the care and feeding of the plumbing necessary for the entire thing to exist and run. As application developers, we do not need to worry about the *ping, power, and pipe* necessary for our application. The cloud provider does this for us with a well-defined **service-level agreement** (**SLA**). The other benefit of this is a little less obvious: when we start using cloud-native technologies and services, our applications are more agile and can respond to unforeseen real-time events in a timelier manner; they can also adapt to change, which always happens, better over time. In short, our applications are better for the modern world we live in. Monolithic applications still have a place, but less and less in this more rapidly changing world.

There is nothing stopping you from deploying a monolithic application into Microsoft Azure. In fact, if you want to do the closest thing to a *lift and shift*, to get off the mainframe or other legacy system as fast as possible, this might be a really good approach. You can migrate your monolithic workloads into Azure and then modernize them after they are there into cloud-native services such as Azure Kubernetes Service (AKS) or Azure Functions. In fact, this is what many organizations are doing. They are migrating and then modernizing.

How do microservices fit into the picture?

So, we touched on microservices earlier in this chapter in the subsection titled *The database-centric nature of a monolith*, but now, let's focus on how they differ from mainframes. With a microservice, the intent is to isolate the processing and data by some real-world association. Typically, one of the most used is **business domains**, which can be agency or business departments. Or, we can isolate them by functionality—whichever makes the most sense. There are other ways we can isolate the services as well. This is one of the more important activities in redesigning and decomposing a monolith.

Let's address the word *decompose*. What do we mean when we say that? What is meant is that we are separating out from the monolith the processing and data associated with a particular service. Keep in mind here that the services in microservices have their own processing and data. So, we are decomposing the applications and the database. When we do this, we decompose the monolith. There are a number of ways we can do this. The most popular approaches are to decompose by the following:

- Business functionality

- Business domain

- Transaction efficiency

The following diagram shows an example of how microservices can be grouped together and called from other microservices. This grouping can be in the form of business domains or functional domains. That decision is up to you and the architects. Contrast this with the monolith from earlier in this chapter where all of those domains were in the same deployment, and you can see how microservices are more agile:

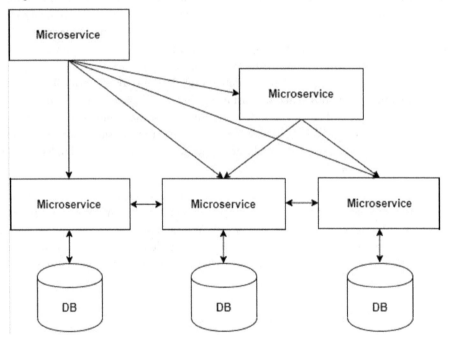

Figure 7.4 – Microservices calling separate microservices with distinct databases

So, now that we know what decomposing a monolith means, let's look at some typical approaches and patterns for real-world scenarios.

Decomposition patterns

A deep dive into **decomposition patterns** is beyond the scope of this book. If you want a book that goes into this topic in depth, I highly recommend Sam Newman's *Monolith to Microservices*. Having said that, let's take a look at an example you have probably seen before—a complex transaction within a database.

In a monolith, we can enlist the resources (that is, tables, views, and stored procedures) that we need to implement a consistent state change, and we can do this under the context of a transaction. The database will make sure for us that either the whole transaction will complete or, if something goes wrong, no state change will occur. This means that if we get to the end of a complex transaction and something goes wrong, the database will roll back the necessary changes to make sure the database is consistent and at the state it was before the transaction. This is known as an **Atomicity, Consistency, Isolation, and Durability** (**ACID**) transaction and is one of the main reasons databases are so reliable and necessary for modern computing.

Now, how would we approach this in a microservices world? Think about it. The database is not a holistic entity anymore. In fact, it is separated and isolated into potentially many distinct databases that can typically only be accessed via the corresponding service. The answer to this question is we would use what is known as a **Saga pattern**. Think *saga* as in a long story; it's not an acronym. Essentially, what we need is a way to call separate services in an orchestrated way and have them initiate the state change to their respective databases. We also need to have the ability to compensate for the state change if one of the services is not able to update its database. So, what we have is a way to have ACID-like transaction management similar to a relational database but in a microservices world.

This is just one example; there are many others, but hopefully, you get the idea. So, what are the benefits of moving to a microservices architecture? There are a number of them; let's look at the most obvious:

- Since each service is its own entity and isolated, I can update just that service and avoid a *waterfall*-type update where I need to update the entire application

- The services can scale independently, thus saving processing costs

- The services can be built using different technology stacks and languages, such as .NET, Java, and even COBOL as a language

So, in a cloud-native world, you can see that adopting microservices can be advantageous. You might be wondering at this point, since I just pointed out that you can even leverage COBOL in microservices, is there a middle ground when moving to cloud native?

One thing to consider about legacy applications, especially ones that have been running for a long time performing mission-critical functionality, is that if the program is working and performing a function that is not going to change, do we really need to rewrite it? It is possible to compile COBOL into .NET and Java, so it is possible, in fact, it is done quite often, to deploy the COBOL functionality into a microservices cloud-native technology such as containers or Azure Functions and have that legacy code participate in the new world order of microservices. So, there is a middle ground when it comes to migration of those workloads to cloud native in Azure.

We've now looked at decomposition patterns. Next, we will summarize the chapter.

Summary

In this chapter, we focused on the monolithic architectures prevalent in mainframes and midrange computers. We looked at what their advantages and disadvantages were and why applications that are born in the cloud and leverage cloud-native technologies are typically not monoliths. We also looked at approaches for decomposing a monolithic application into microservices and the middle ground for moving to cloud native when adopting microservices.

Now that we have explored how to bridge the gap in architectures, let's focus on the deployment options available when we migrate legacy workloads to Azure in the next chapter.

Part 3:
Azure Deployment and Future Considerations

This section will look at deploying your legacy application to Azure in an optimal manner, and how to take advantage of new Azure features as they become available.

This part contains the following chapters:

8
Exploring Deployment Options in Azure

When it comes to legacy workload deployment options, they were pretty specific to the platform you were using at the time. This is especially true for mainframes where there was a well-defined and accepted way to deploy applications. There was not a lot of variety in transaction managers, filesystems, queueing architectures, and databases. You used what was on the machine platform, and whichever tool you decided to use, it usually did a very good job at whatever it was supposed to do. A good example of this is **Customer Information Control System** (**CICS**) on IBM mainframes. If you were writing online applications on a mainframe, this is what you used to communicate with the database (typically Db2) and the user (typically a 3270 terminal). IBM midrange systems were very similar. Legacy Unix systems did allow more variability but, for the most part, had a specific set of vendor solutions that did a specific job for your application.

In the world of cloud hyperscalers, however, we have a lot more options for deployment. Each option has a particular advantage, depending on what you are designing for. In this chapter, we are going to examine the most popular deployment options since they will probably be the ones you will end up using in some form or fashion.

The chapter covers the following topics:

- What exactly do we mean by deployment options?
- Good ol' **virtual machines** (**VMs**)
- **Azure Service Fabric** (**ASF**)
- Azure Functions
- Containers
- Azure App Service
- Azure Stack

What exactly do we mean by deployment options?

When we are referring to *deployment options*, we are specifically talking about how your legacy application is deployed in Azure. Since Azure is a public cloud platform, there are more options to choose from, depending on how you want the application to behave. These options run the spectrum from traditional deployment options, such as VMs, to more cloud native, such as ASF. In this chapter, we will explore the main options you will have available and the benefits and considerations of each, so let's have a closer look at these:

- **VMs**: If you are reading this book coming from an on-premises environment, you would have spent the last 20 years under a rock to not know what this option is. Let's focus on the ones you might not have heard of. More on this later in the *Good ol' VMs* section.

- **ASF**: ASF is a serverless deployment option available in Azure. I have to admit, I am a big fan of ASF, mainly because it allows you to create cloud-scale applications that inherit cloud capabilities just because of their deployment option. If you have ever heard of or used Azure SQL Database or Azure Cosmos DB, then you'll have an idea of the types of applications that can be built with ASF. It is a deployment option for containers and microservices that comes with a lot of benefits.

- **Azure Functions**: Azure Functions is another serverless deployment option for functionality that is typically event-driven. It is scalable and available and allows you to respond to real-world events.

- **Containers**: Containers, as you might already know, are a very popular deployment option that has been around for a while. They are similar to VMs in that they allow you to have a lot of control over the application environment, but allow you to abstract the OS away from your deployment and only focus on the necessary services to support your application. They can be deployed into a number of Azure services, which we will dive into in just a bit.

- **Azure App Service**: When you need an application environment that just allows your application to execute, you might like Azure App Service. Great for Tomcat or Apache-type applications that just need an environment to run.

- **Azure Stack**: Azure Stack is a little different in that it is a deployment option of Azure that can be local to your on-premises environment. Imagine an Azure region in your on-premises data center. That is what Azure Stack and its various forms are. It is, however, not a full implementation of Azure; there is always a catch.

So, there you have it. These are the options that we are going to look at. Are there others? The answer is yes, but not ones that you will probably use for mainframe legacy applications. As we evaluate these options, we are going to look at the pros and cons and why you would use one over the other. Keep in mind, as we stated earlier, that these deployment options are not mutually exclusive, and in some cases, you might choose one just based on preference. Having said that, let's take a look at the old kid on the block—VMs.

Good ol'VMs

In the case of a rehost or refactor of the mainframe workloads, you will quite often need to consider VMs. This is the oldest and probably most common deployment option. Not because it is better, but because it has been around longer than any of the other options we are going to consider. Even before distributed platform virtualization was available, mainframe workloads were offloaded by various methods onto physical machines. It was quite logical for this model to incorporate VMs when that technology became available. Let's have a look at the benefits and things to consider when deciding on this option.

Benefits

Let's look at the benefits of VMs:

- As we mentioned, the solutions available by mainframe rehosting and refactoring vendors—with the exception of a few—all run on VMs. This is because most IT shops are very familiar with the care and feeding of this deployment model. They are reliable and, with modern tooling, easy to administer and manage. As in previous chapters, we have talked about how we can make VMs highly available and resilient, so there are a lot of technologies that can be applied to VMs, which ensures their enterprise-class standing.

- With technologies such as Azure Scale Sets, VMs can be elastic and scale up or down. So, in some ways, they can achieve similar benefits to containers. They can also be scaled out (that is, made into larger VMs), although this usually requires downtime. With Azure features such as Azure Site Recovery, you can add **business continuity** (**BC**) and data recovery to VM implementations and make it highly available across Azure Regions.

Things to consider

Now let's look at things you will need to consider when using them:

- Probably the single biggest thing to consider with the VM option is that you need to provide and care for the OS, whether it is Windows or any flavor of Linux. Now, this can be considered a good thing when it comes to flexibility, but with that flexibility comes complexity. You will have to deal with system drivers, configurations, and things such as that. And don't forget updates and patches. Does this bring back memories? This is why containers are becoming more and more popular.

- Mainframes can scale using a feature called Capacity on Demand. With Azure we can use an even more automated technology such as Azure Scale Sets, VMs can be elastic and scale up or down. They can also be scaled out (that is, made into larger VMs), although this usually requires downtime.

- Mainframes are typically scale-up, monolithic environments. This is why VMs; as we mentioned, the solutions available by mainframe rehosting and refactoring vendors—with the exception of a few—all run on VMs. This is because most IT shops are very familiar with the care and feeding of this deployment model. They are reliable and, with modern tooling, easy to manage. Having said that, the current trend in IT is to lean more toward automation, and, let's face it, there are other deployment options that lend themselves to this end much better.

ASF

ASF is a deployment option that can be very compelling, especially when your goal is to leverage cloud-native services that exploit the benefits of a modern cloud architecture such as Azure. Where technologies such as Kubernetes excel at typically stateless applications, ASF has a strong focus on stateful applications. What exactly does this mean? Stateless applications execute and do what they are supposed to do and then go away. They literally *go away* in the digital sense in that they have no idea of what they were doing before unless you maintain the state of what they were doing in some place such as a Redis cache. Stateful applications, on the other hand, do maintain their own state. Depending on the application you are creating, one will be better suited than the other. The other thing to keep in mind about ASF is that an ASF cluster can be deployed on-premises or on another cloud. Think about that. It offers most of the benefits of Kubernetes and can span Azure from on-premises to another hyperscaler. You can orchestrate both stateless and stateful containers through different deployment locations and different phases of the life-cycle management of the application.

Keep in mind that this is a native Azure service, so there's no need to care for and feed it for maintenance. Now, let's look at the pros and cons.

Benefits

Here are the benefits of choosing ASF:

- The benefits of ASF are pretty obvious when you are targeting a modern cloud-native deployment option. You get most of the benefits of a platform such as **Azure Kubernetes Service** (**AKS**), and it allows for the benefits we expect from a cloud service, such as scalability and reliability.

- Since ASF is based on containers, it offers yet another deployment option for the very popular container packaging of applications. This means you can choose between AKS, which we will talk about later, or ASF.

Things to consider

There are, of course, some things you will need to consider:

- So, what is the catch? Well, the catch is getting there. As we will find out with some of the other deployment options, modernizing mainframe applications to a modern application deployment option such as ASF requires considerably more effort than, say, VMs. Monolithic applications are often described as being *chatty*, and the reason for this is that they *could* be chatty. They were created on monoliths and did not need to worry about traversing a network to talk to other applications like distributed applications do. When you take that logic and approach to application design, you need to step back and consider your approach.

- For certain scenarios for refactoring and rehosting, ASF is a viable deployment option. As a matter of fact, there is a case study of a relatively well-known automobile manufacturer where ASF was used in a mainframe modernization effort, and it worked flawlessly. Having said that, if you have a chatty application that communicates with other artifacts, you will need to consider how to leverage ASF in the most effective way for your modernization project.

- Keep in mind that ASF is built on top of AKS. As we mentioned, there are cloud-native services that are built using ASF. Cosmos DB is a good example of that. As a result, it is a complex deployment model that is really only suitable for certain types of workloads.

Many legacy workloads were predicated on the concept of online versus batch processing. As we touched on previously, in *Chapter 2*, this meant that during operating hours, employees were working on the computer entering transactions, and either during or after hours, batch processes ran to process the day's transactions. In the new world order of cloud computing, the concept of online versus batch is clouded, to say the least… pun intended.

What has changed? The fundamental difference is that instead of being reliant on nightly batch jobs to calculate what happened during the day, now we can process those transactions as they happen with eventing technologies. This is the next deployment option that we will look at—Azure Functions.

Azure Functions

Azure Functions allows you to deploy code that executes when certain events happen. That might sound kind of lame and generic, but when you consider that we live in an **Internet of Things (IoT)** world now, it might make more sense. The ability to respond to and process events as they happen leads to a more real-time IT world where the need for batch processes does not necessarily go away but is greatly diminished. Azure Functions is a deployment option that can be leveraged for this paradigm shift. Think of Azure Functions as *compute on-demand*.

This is somewhat of a specific and boutique workload in that it can only be used for certain workloads and usually in a re-architecturing and reengineering effort, but nonetheless is a viable deployment option when you are modernizing mainframe workloads.

Benefits

Now that you have an idea of what Azure Functions is, let's look at the benefits and things you will need to consider if you choose to select it for your deployment:

- Primarily, Azure Functions can be used in the context of batch processing of migrated legacy workloads. When you think about these types of workloads, things such as file uploads, scheduled tasks, events in databases, or queues come to mind.

- We should probably mention that since Azure Functions is offered as a **cloud-native service**, which means you simply need to deploy your application code and start using it. Your code needs to be simple and respond to certain and very specific scenarios, which is another benefit. This has another added benefit in that it allows you to maximize your compute spend and minimize your deployment packages. In other words, you can deploy very lightweight packages that utilize minimal compute and respond to real-world events as needed.

Things to consider

Here is a caveat you will need to be aware of when using Azure Functions:

- A warning! When you really start to consider Azure Functions in your modernization journey, it will make you re-evaluate your thinking about batch processing. Be forewarned. The need for nightly batch processing essentially starts to fade away, and you start to enter a new processing model where you respond to events as they happen. In the end, this is better, but as we all know, there will probably always be a need to do some form of batch processing, and there are provisions within Azure, such as Azure Batch.

Now that we have taken a look at some very specific Azure technologies, let's broaden the scope and look at containers.

Containers

As with VMs, you would have to have been under a rock for the last 10 years to not know what a container is. However, if you are coming from a legacy environment, you might have heard about them but not really worked with them. Let's do a very quick refresh.

Essentially, containers let application developers break free from the prison of VMs. What do I mean when I say this? Consider that in the *old days*, you had to deploy your application or applications on a VM. That was it. You might have been forced to deploy multiple applications on one VM or maybe just one. With that, what happened when you needed to update the OS? Did that update affect all applications the same way? This is just one scenario of what happened when the VM model started breaking down and thus came in the age of containers. Containers solve the aforementioned problem and others as well. Too many to go into. The bottom line is that for modern applications, containers are here to stay for a while, and this is a good thing.

So, since containers are the ever-rising star in the field of application deployment, Microsoft Azure provides a variety of deployment options for them. Let's have a look:

- **AKS**: Containers are great, but in order to use them in an effective way, you need to orchestrate them based on the real-world scenario you are solving. Kubernetes has addressed this need for a while now, but you need to install and configure the services that make up your Kubernetes cluster. AKS is a fully managed service that manages your Kubernetes cluster. It takes care of critical tasks such as maintenance and health monitoring. You tell AKS the number and size of the nodes that make up the cluster, and AKS deploys and configures the deployment. You literally then just need to bring your containers, and you are off to the races. It greatly simplifies the administration and management of Kubernetes.

- **Azure Container Instances** (**ACI**): ACI is similar to AKS but is targeted toward workloads that do not require complex orchestration. It provides a very easy deployment model for containers and completely eliminates the need to manage VMs. It is based on a serverless cloud-native model. When you deploy an AKS cluster, you still need to decide what type of VM the cluster will be built on. With ACI, you do not need to do this, so in a sense, it is more cloud native in that capacity. When you need a more lightweight deployment option than AKS, ACI could be the deployment option you are looking for. Another thing to consider is that ACI can easily integrate with AKS, so they are not mutually exclusive. ACI can also integrate with Azure Logic Apps, which opens up a new area of cloud application development.

- **ASF**: We have already looked at ASF, which is another deployment option for container-based applications. You will need to consider this if you are dealing with more stateful containers.

- **Web App for Containers**: Azure Web App for Containers is technically a part of what is known collectively as **Azure App Service**. Azure App Service is an Azure set of services targeted toward running web applications in general. They can be Windows based or Linux based. App Service is great for running your presentation tier. Web App for Containers is a specialized option of Azure App Service that is focused on containers. In the scheme of complexity and control, you can think of Web App for Containers as being useful for even more lightweight deployments of containers.

- **Azure Spring Apps**: The Spring Framework is gaining popularity. It is a framework for application development based on Java and supports containers as well. If you have an investment in the Spring Framework or have decided that this might be a fitting deployment option for your migrated workloads, you will need to look at Azure Spring Apps. This allows you to run Spring Boot, Spring Cloud, and other Spring applications in Azure. As you would think, it is a fully managed service that integrates with the other deployment options we have been examining.

- **Azure Red Hat OpenShift**: Much like the Spring Framework, another deployment option to consider is Azure Red Hat OpenShift. I have found that the best way to think of OpenShift is as a specialized implementation of Kubernetes. They each have their pros and cons, which are beyond the scope of this book. Kubernetes is generally more flexible than OpenShift. OpenShift is targeted toward enterprise applications. They both excel at orchestrating containers but have a slightly different approach. As the name implies, OpenShift is an offering that is jointly engineered, supported, and operated by Red Hat and Microsoft, whereas Kubernetes is an open source initiative. If you are already using OpenShift or are considering it for your cloud modernization journey, Azure Red Hat OpenShift could be a compelling option for your deployment.

So, there you have it. As you can see, when it comes to deployment options for container-based applications, you have quite a few options. Each has its strengths and weaknesses, depending on the types of workloads you are deploying. One last option that we need to look at—we have already looked at it briefly, but it warrants a deeper look—is Azure App Service.

Azure App Service

As we mentioned, we already looked at the component of Azure App Service that allows you to run containers, Azure Web App for Containers. Azure App Service also allows you to run non-container-based applications as well. It is a great service for running .NET, .NET Core, Node.js, Java, Python, and PHP web-based applications. For any given legacy installation, there is typically a service facade built around it on-premises that supports things such as external access to CICS transactions, a web-based UI for interacting with Db2 or **Virtual Storage Access Method** (**VSAM**), and also applications that need to integrate with the mainframe and/or legacy estate. These types of applications are good candidates for Azure App Service. As you would guess, it is fully managed and cloud native. You do not need to worry about standing up the VM and OS. Azure handles all of that for you. You literally bring your applications and they just run. It has built-in scalability and resilience both inside the Azure Region you are deploying in and across Azure Regions.

Benefits

Now, let's look at one benefit of Azure App Service:

- The main benefit of Azure App Service is, obviously, its ease of use and deployment. It integrates with other Azure services and, of course, the Azure security services, so it is secure.

Things to consider

Here is one thing you will need to consider when using Azure App Service:

- The single biggest concern for Azure App Service is what is typical for any managed service, and that is you sacrifice a level of control. If you have a highly customized managed application environment, you will need to look at the limitations of Azure App Service. But, if your applications are pretty standard, you should be okay.

There is one more deployment option we need to look at: Azure Stack. Azure Stack allows you to bring a part of Azure to your on-premises data center. Let's have a closer look.

Azure Stack

One of the most consistent issues that you will need to consider when you are modernizing your legacy workloads is **network latency**. The latency between the on-premises data center and the Azure Region can sometimes be unacceptable even when using a technology such as **ExpressRoute**, which we looked at earlier in the book in *Chapter 6*. When you encounter this issue, you will need to look at Azure Stack. Azure Stack contains many of the essential Azure resources (especially IaaS) that you will need in an on-premises deployable instance. It is offered by hardware vendors and provides many of the benefits of Azure, but not all of them. You will need to see which services you intend to use to make sure that Azure Stack is the solution you need. As of the writing of this book, it comes in three flavors. Let's take a look at them:

- **Azure Stack HCI**: Azure Stack HCI is a **hyper-converged infrastructure** (HCI) for running **Platform as a Service** (PaaS)- and IaaS-based workloads on-premises. You can run products such as Microsoft Exchange, SharePoint, SQL Server, and even AKS. Azure Stack HCI can integrate with cloud-based Azure, but it can also operate independently from Azure as well. It is typically used for these types of scenarios:

  - Back offices

  - Virtual desktop integration

  - Scalable storage

  - High-performance database scenarios

- **Azure Stack Edge**: This offering of Azure Stack is not really relevant for legacy modernization unless you intend on leveraging IoT-type scenarios.

- **Azure Stack Hub**: Azure Stack Hub will actually be very relevant for your modernization journey. This is one of its target use cases. It lets you run apps in an on-premises environment and deliver Azure services in your data center. If you need to have Azure core services such as VMs and Kubernetes services, Azure Stack Hub can provide those types of services to remote locations that can be disconnected with intermittent connectivity to the public Azure cloud. This is great when you need to process data locally and perhaps need it close to the legacy installation to avoid latency issues. Keep in mind this is a subset of Azure and does not have many of the cloud-native services we have been exploring in this book. Another problem Azure Stack Hub solves is that of compliance and data sovereignty. If you need to comply with governmental or industry requirements, this is an option to consider.

Summary

So, this brings us to the end of our overview of deployment options in Azure. Hopefully, this was useful and will help you have a better understanding of the benefits and concerns of each type of deployment option. One thing you probably noticed is that there is a lot of overlap between some of these services. You will need to consider other things when deciding on which option to use, such as what is the culture of development staff and how agile and flexible your modernized application needs to be. In the next chapter, we will explore modernization strategies at a high level. We will look at the patterns and best practices for running your legacy applications in Azure.

9
Modernization Strategies and Patterns – the Big Picture

The strategy, or strategies, available for you to modernize a legacy system to Azure will largely depend on your goals for running the system you wish to migrate. We will discuss in detail the following four migration goals, and what strategies might work best for each type of goal:

- Exiting the existing data center as quickly as possible
- Taking a multi-step (staged) approach
- Modernizing to a cloud-native infrastructure
- Redesigning the system for Azure

Each category will likely have more than one modernization strategy available. It's up to you to pick the best approach for your situation.

This chapter will cover the following topics:

- Data center exit – how to get to Azure as quickly as possible
- Redesigning for Azure – completely changing the application
- Staged modernization strategy – taking a multi-step approach

Let's start with the requirement to exit the data center as quickly as possible.

Data center exit – how to get to Azure as quickly as possible

There could be a number of reasons why a legacy application needs to modernize to Azure as quickly as possible, including the following:

- The customer needs to vacate the facility that houses the legacy system
- The legacy system is the last application remaining in the data center, driving up costs
- The contracts on the legacy hardware and software have a pending renewal that will increase costs

The key factors all of these reasons have in common are as follows:

- Costs related to keeping the data center running are too expensive
- The application itself currently meets business requirements

This implies that minimal code changes are required. However, the target architecture on Azure and deployment methodologies can be adapted to the needs of a particular customer.

Moving to emulators on VMs

Emulators are probably the fastest way, with the least risk, of moving applications from a data center to Azure. Just to clarify terms, for the purpose of this book, the term **emulator** means an application can run *as is* with no recompiling on the target VM running. With that said, not all legacy solutions have emulation options. The following are some emulators that are available for legacy systems:

- **Solaris**: Using third-party **independent software vendor (ISV)** tools
- **HP-UX (PA-RISC)**: Using third-party tools
- **Unisys Libra on Azure**: Tools from Unisys
- **IBM zD&T**: For z/OS development and test only (not production)

When all is said and done, an application running on an emulator in Azure is really no different than any other application running on Azure. You need to consider compute, storage, and network requirements for the application, plus the non-functional requirements such as **service level agreements (SLAs)**, latency, and **high availability/disaster recovery (HA/DR)**. The following are a number of the things to consider when hosting a VM-based solution in Azure:

- **Compute required**: Keep in mind that emulation adds some overhead. Typically, we see a 10 to 20% overhead for emulation. This would be in addition to the compute estimated from the current legacy platform.
- **Storage, both IOPS and throughput**: We find optimizing storage to be a key success factor.

- **Network requirements**: While Azure data centers likely have the network throughput needed within them, any connections to existing systems on premises should also be addressed.

- **High availability**: This can include failover capabilities native to the emulator operating systems, or using redundancy, as with Azure Load Balancing.

- **Security**: While the emulation likely provides security similar to the legacy operating environment within the VM itself, access to the VM will need to use Azure services such as Azure Active Directory, Azure vNets, Bastion, certificates, and so on.

To organize these preceding items, we recommend using an **Azure Landing Zone** (**ALZ**) template to simplify deployment.

Finally, when deploying emulation to VMs, some ISVs require specific settings for the compute services, such as turning off (if needed) hyper-threading.

The following diagram summarizes the typical features needed to deploy a VM using third-party emulation:

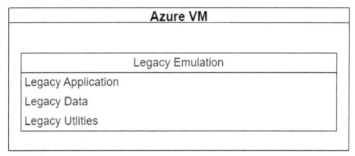

Figure 9.1 – VM emulation

Now that we've looked at deploying VMs using emulation, we will next look at deploying ISV tools that do not require operating system emulation.

Deploying the VMs that use ISV tools

It's important to differentiate between emulators and tools that map APIs (such as `EXEC CICS` statements) to functions developed natively for x86 that run on Azure. Solutions using these ISV tools are even more like deploying an x86 VM on Azure. That said, these solutions are similar to deploying emulators on a VM. The biggest difference is the integration with Azure services such as Azure Active Directory, Azure Redis Service, and Azure Load Balancing.

These ISV tools are typically for either IBM or Unisys mainframes with COBOL or PL/I applications. The type of APIs these tools support include the following:

- IBM CICS

- IBM IMS TM

- IBM Db2

- IBM IMS DB

- Utilities for IBM mainframes for scheduling, data movement, and administration

- Unisys MCS

- Unisys TIP

- Unisys DMS

- iSeries data and utilities

These types of tools will usually include options for how data is deployed. For example, a COBOL programmer may use mainframe Db2 syntax, but the ISV tool converts that syntax to SQL Server. Therefore, the deployment will be on the target database (such as SQL Server), and not necessarily on Db2 **LUW** (**Linux, UNIX, Windows**). That said, Db2 LUW may also be an option, depending on the ISV tool.

When all is said and done, an application using tools that support mainframe syntax in Azure is really no different than any other application running on Azure. You need to consider compute, storage, and network requirements for the application, plus the non-functional requirements such as SLAs, latency, and HA/DR. The following are a number of the things to consider when hosting a VM-based solution in Azure:

- **Compute required**: This will vary a bit, depending on the ISV tool and what language is used (migrating Assembler typically takes more compute). That said, a conservative estimate is to allow for 1 vCPU in Azure for each 150 **million instructions per second** (**MIPS**) of an online application, and 1 vCPU per 100 MIPS for a batch application.

- **Storage, both IOPS and throughput**: We find optimizing storage to be a key success factor.

- **Network requirements**: While Azure data centers likely have the network throughput needed within them, any connections to existing systems on premises should also be addressed.

- **High availability**: This can include failover capabilities with multiple VMs, or using redundancy, as with Azure Load Balancing.

- **Security**: While ISV tools often provide their own layer of security similar to the legacy operating environment within the VM itself, access to the VM will need to use Azure services such as Azure Active Directory, Azure vNets, Bastion, certificates, and so on.

To organize these preceding items, we recommend using an ALZ template to simplify deployment.

The following diagram summarizes the typical features needed to deploy a VM using third-party emulation ISV tools:

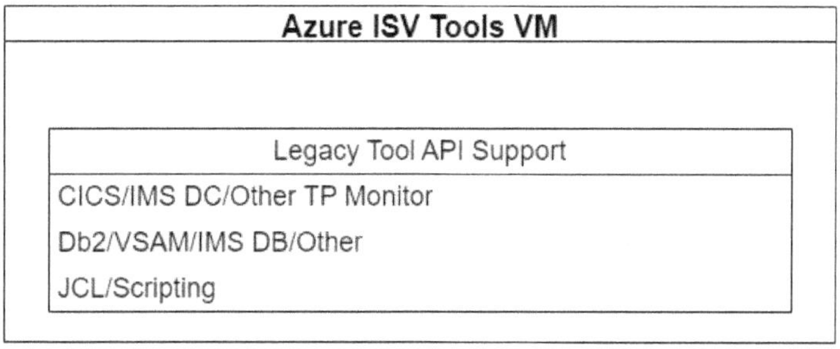

Azure ISV Tools VM

Legacy Tool API Support
CICS/IMS DC/Other TP Monitor
Db2/VSAM/IMS DB/Other
JCL/Scripting

Figure 9.2 – VM ISV tools diagram

Now that we've looked at deploying VMs using ISV tools, we will next look at deploying the migrated applications to native Azure tools.

Deploying to containers and PaaS

In addition to **virtual machines (VMs)**, PaaS services on Azure, such as **Azure Kubernetes Service (AKS)**, Azure SQL **Managed Instance (MI)**, and Azure application services such as Service Bus (for queuing) and Azure Event Hubs (as a scheduling alternative) may be options. Let's look at these options in three categories:

- Container/application services
- Data services
- Utility services

Container/application services

Both the rehost (keeping the same language) and refactor (changing the language but not the architecture) will often have the option for deploying to containers, such as AKS or Azure Container Instances, or to Azure Application Service. The choice largely depends on the needs of the application that is being migrated. For example, if the application is predominantly an online system, then Azure Application Services might be the right choice. If the application needs to auto-scale depending on load, then AKS might be the right choice.

Regarding compute sizing, deploying container/application services is very similar to deploying ISV tools on VMs, as covered in the prior section. The one exception is to expect an approximate extra 10% overhead for these tools. Based on this, you should allow for 1 vCPU per 150 MIPS for online, and 1 vCPU per 100 MIPS for batch. Additionally, however, you should increase each total by 10%. For example, if you estimate 22 vCPUs for a 3,300 MIPS online application, you should increase that total by 2 vCPUs, for a grand total of 24 vCPUs.

Other considerations include the following:

- **Storage, both IOPS and throughput**: This will largely be taken care of by PaaS data services, so only additional storage such as Blob or files is typically required.

- **Network requirements**: While Azure data centers likely have the network throughput needed within them, any connections to existing systems on premises should also be addressed.

- **High availability**: This can include failover capabilities. Container services such as AKS have options for both local and regional redundancy.

- **Security**: While ISV tools often provide their own layer of security similar to the legacy operating environment within the container service itself, access to the container service will need to use Azure services such as Azure Active Directory, Azure vNets, certificates, and so on.

Data services

Azure offers a wide variety of PaaS data services, including the following:

- Azure SQL DB (relational)

- Azure SQL MI (relational)

- Azure PostgreSQL DB (relational)

- Azure MariaDB (relational)

- Azure MySQL DB (relational)

- Azure Cosmos DB for NoSQL

- **Azure Data Lake (ADL)**

- Azure Synapse (massively parallel database)

The majority of legacy applications that choose to be deployed on PaaS data services will likely target either a relational service or NoSQL.

Cosmos DB is the NoSQL option for Azure. This might be a great choice for moving non-relational data, such as **virtual storage access method (VSAM)**, to Azure. However, since there are several choices for relational databases, here are some general guidelines for choosing the right relational data option:

- **Azure SQL DB**: For native Azure implementations that do not require specific compatibility with on-premises features of SQL Server

- **Azure SQL MI**: For systems that need compatibility with an on-premises SQL Server

- **Azure PostgreSQL DB**: When PostgreSQL features are desired

- **Azure MySQL DB**: When MySQL DB features are desired

- **Azure MariaDB**: When MariaDB features are desired

As with containers and application services, there are several decision areas that need to be addressed when deploying to Azure:

- **Compute sizing**: Each option provides a compute sizing option. The general guidance for sizing relational databases on Azure is to allocate 1 vCPU per 100 MIPs for the legacy system.

- **Storage**: Each data PaaS option allows the customer to size the data and IOPS: both IOPS and throughput.

- **Network requirements**: While Azure data centers likely have the network throughput needed within them, any connections to existing systems on premises should also be addressed.

- **High availability**: Each type of PaaS database has an offering for HA.

- **Security**: While each PaaS database has its own layer of security, access to the PaaS database will need to use Azure services such as Azure Active Directory, Azure vNets, certificates, and so on.

Utility services

Legacy applications often use a number of utility services such as queuing, eventing, scheduling, virtual tape, data import/export, and other services. On Azure, most of these capabilities can be mapped to Azure PaaS services as follows:

- **Data import/export**: **Azure Data Factory** (**ADF**)

- **Queuing**: Azure Service Bus

- **Eventing**: Azure Event Hubs

- **Scheduling**: Azure Scheduler

- **Virtual Tape**: Azure Backup and Blob Storage

Sizing these services usually depends on the volume of transactions or throughput needed. Also, some of the services offer options to work over multiple regions. It is important to understand your application's requirements to determine the right size and SLA for each service.

In this section, we discussed the various types of PaaS services an application might use when moving from a legacy platform to Azure. The following diagram summarizes these services:

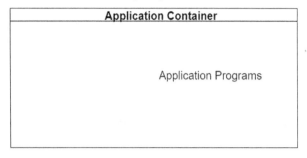

Figure 9.3 – PaaS

Now that we've looked at considerations for deploying to PaaS services, next, we will look at deployment when changing a legacy application to a new Azure native application.

Redesigning for Azure – completely changing the application

There might also be compelling reasons to completely redesign the application to native Azure when moving off a legacy system. The following are a number of potential reasons:

- The system does not provide the business functionality needed on the current legacy platform. In this case, it would not make sense to move the system *as is* to Azure.

- The current legacy system has a lot of **technical debt**, meaning that even if the current functionality meets business requirements, your company does not have the trained staff to support or modify the system.

- You want to take advantage of cloud-native features, such as Spark compute clusters, artificial intelligence, or on-demand scaling.

The key things you need to consider before undertaking a project that does a complete redesign include the following:

- Understand the business functionality: what you want to keep, what you want to drop, and what you want to change

- Have the technical knowledge available for the Azure-native services you want to target

- Build the test plans you need to validate both functional requirements and non-functional requirements

- Understand both the scope and effort involved to deliver the redesigned system

While redesigning the application to be Azure-native may take the most effort and time, our experience is that customers are usually happiest with the outcome.

The following diagram shows examples of cloud-native Azure services that can be used for both applications and data:

Application Services

Microservice based
Deployed as Azure Functions
Spark-based analytics for scale

Data Services

SQL DB for relational data
Cosmos DB for NoSQL data
Azure Data Lake for analytics storage

PaaS Utility Services

Event Hubs for scheduling
Service Bus transaction durability and resilience
Azure Databricks for Spark clusters

Figure 9.4 – Redesign example

Now, we will look at using two or more of these options combined.

Staged modernization strategy – taking a multi-step approach

While you may ultimately want modernization, you may find that there are goals you need to meet prior to the time needed to make that type of transaction. For example, you might need to get out of your current data center in one year, and your most optimistic estimate for modernizing your application to cloud-native will take a minimum of two years to complete.

In addition, you may find that some of the applications you need to modernize will get less benefit from a complete transformation to Azure-native than others. In this case, you may choose to rehost or refactor some applications first and concentrate on other applications that benefit most from a complete transformation to Azure-native.

In either case, as described previously, you could benefit from a multi-step approach for modernization to Azure. Taking a multi-step approach can also allow your modernization staff to stay focused on fewer applications at a time. Here are some of the advantages of a multi-step approach:

- Get out of the existing data center quicker

- Achieve benefits from having the applications and data quicker in Azure

- Fewer resources are required to complete the modernization

- The ability to set up a *migration factory* to handle multiple modernizations

Next, we will cover the preceding fourth item, a modernization factory, in more detail.

Creating a modernization factory approach

A **modernization factory approach** is where you have multiple legacy applications you want to move to Azure, and you set up a repeatable process to address this. Some of the features addressed by a modernization factory are listed next, followed by another list of the advantages.

The tasks handled by a modernization factory are as follows:

- Assess application

- Create target architecture on Azure

- Triage application for the migration effort

- Run any proof points needed for the application

- Convert code

- Convert data

- Address integration points

- Functional tests

- Performance tests

- Acceptance tests

The advantages of a modernization factory are as follows:

- Repeatable process

- Standardization

- Increased quality

- Faster completion time

- Allows coordination across teams

While a modernization factory is overkill for a single system or a small system, the advantages when there are multiple systems to move to Azure easily justify the effort and time to set up the modernization approach.

Also, while we are discussing the migration factory approach in the multi-stage section of this chapter, a migration factory can apply to any of the methods we discussed in this chapter.

Now, we'll move on to the summary.

Summary

To summarize, the following table shows the description, advantages, disadvantages, and likely application candidates for each modernization approach to Azure discussed in this chapter:

Approach	Advantages	Disadvantages	Candidate System
Legacy Emulation	Fastest to Azure Fewest technical risks	Not available for all legacy systems Does not modernize application	Application that is not changing Application that will be retired soon Application runs on a platform that supports emulation

Rehost with Partner Tools	Fast time to Azure Low risk Leverages existing skills Moves data to Azure for potential analytics	Still requires legacy language skills Does not address technical department Often requires partner runtime	Application that meets business needs Application has source Customer has personnel to support legacy language
Refactor with Partner Tools	Medium time to Azure Medium risk Allows newer developers to maintain application Addresses some of the technical department Data usable on Azure	Takes longer than rehost Takes longer to test Refactored code quality might be low Typically, costs more than rehost	Customer that wants to get rid of legacy language but wants to keep the functionality with minimal architectural changes
Rewrite System to Azure	Takes the most advantage of Azure features Typically, customers are happiest when a completely rewrite to native Azure takes place	Takes the longest and is the most expensive Has the most risks, and requires the most testing	Customer that wants to take advantage of cloud-native services Current legacy system does not meet business requirements
Multi-Step or Staged Method	Reuses processes and procedures Can reduce time and increase quality	Takes time to set up	Customer with a large number of systems to migrate to Azure

Table 9.1 – Advantages and disadvantages based on the approach

The next chapter will specifically look at best practices for modernizing large, monolithic systems.

10

Modernizing the Monolith - Best Practices

In *Chapter 7*, we explored monolithic architectures and contrasted them with microservices-based architectures. We discovered that they are very different in their approach to interprocess communication, especially when it comes to database access. As a reminder, in monolithic architectures, it is quite common to have one or a few databases with many different applications or services from different business units accessing them. In microservices-based architectures, we want to have every service leverage its own database. So, in theory, in a microservices architecture, there is a database for every service we implement.

In this chapter, we will introduce some approaches for modernizing these legacy monolithic applications. We will also look at the best practices to keep in mind on your legacy modernization journey.

We will cover the following topics in this chapter:

- Understanding how we got here – the evolution of application development
- A deeper dive into microservices
- What is the Saga pattern?
- What is the Strangler Fig pattern?
- Introducing Dapr
- Other essential microservices components
- In modernization, one size does not fit all
- Best practices for modernizing the monolith

Understanding how we got here – the evolution of application development

Before we dive into the best practices for modernizing monolithic applications, let's take a step back and look at how we got here. As you may recall from *Chapter 7*, monolithic application design has a direct relationship with the hardware at the time. The hardware environment was typically a large mainframe computer with communication devices hooked up to it for users and **input/output (I/O)** devices. Applications and their related components were tightly integrated into a single code base. This meant that over time, as certain features and functionality needed to be updated, the entire monolith needed to be recompiled.

Gradually, as applications grew larger and more complex, developers began to break these monoliths into smaller, more manageable components called modules. I remember back in the 1980s when modular application development was all the rage. There was even a language called **Modula-2**, which was a variant of **Pascal**. If you were a developer back then, you might remember that. Modular development was a good thing since it allowed you to break the application up along functional and business lines. The underlying issue at this point was that all of these modules were still running on the same monolithic hardware and typically shared the same database or databases.

It took the advent of distributed computing to speed up this evolution. Gradually, as midrange and PC-based servers took hold, applications were being deployed across various machines and platforms. As this happened, modular development led to **service-oriented architecture (SOA)**. With SOA, the message communication between the various services that comprised the application became extremely important. This gave rise to solutions such as **Enterprise Service Bus**, which allowed services to communicate and interoperate across various platforms inside and outside the organization. SOA can be thought of as a coarse-grained service architecture. The services here were exposed via well-known message formats and the **application programming interface** (API), which abstracted the implementation from the service definition. When one service called another, all that needed to be adhered to was the communication protocol and interface. The services were self-contained and could be implemented as a monolith. Many mainframes had and still have an API services layer to allow applications outside the mainframe to access applications and services on the mainframe.

The next stage in this evolution was the advent of microservices architectures, which took the idea of SOA to its next logical conclusion. A microservices architecture divides an application into several smaller, self-contained services, each of which has a database and API. These services are intended to have a loose coupling between them, which implies that they may be independently created, tested, and deployed without causing any disruption to the operation of the other services.

The following figure shows this evolution from monolithic applications to microservices-based applications:

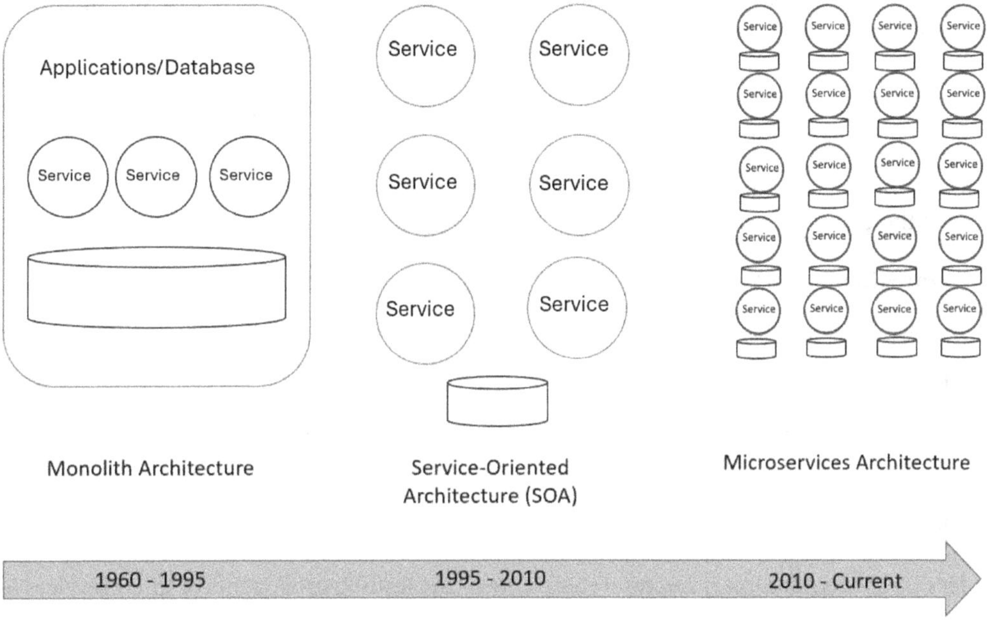

Figure 10.1 – The evolution of application architectures

Microservices offer several advantages, including enhanced scalability, robustness, and adaptability. However, they also present additional issues, such as how to organize updates and deployments, how to manage the complexity of various services, and how to ensure that data is consistent across all of the services. Containerization, orchestration, and service meshes are some examples of the new tools and processes that have arisen in response to the aforementioned difficulties.

A deeper dive into microservices

First, a word about **containers**: before we can take a deeper dive into microservices, we need to have a good understanding of the building blocks that make up microservices, which are containers. Essentially, containers are lightweight, standalone packages of executable code. The container includes everything necessary to run the application code within it: things such as configurations, code libraries, and other code dependencies. The container is void of an operating system, and as such, it is lightweight and allows developers to consistently package and deploy applications in a repeatable way. This makes it very easy to deploy these container-based applications to different environments, such as production or development.

Containers differ from **virtual machines** (**VMs**) in that, as mentioned earlier, they do not have to worry about the care and feeding of an operating system. They are housed in a host operating system that handles those duties. Typically, Linux or Windows are the host operating systems. Since containers are housed within an operating system, they can rely on the operating system to handle the core services and application plumbing, which means that they are very fast and more agile than a virtual or physical machine.

Containers can be housed in a variety of different ways. You might recall that we provided an overview of some of these in *Chapter 8*. Here is a brief recap:

- **Azure Kubernetes Service** (**AKS**): A fully managed Kubernetes service in Azure

- **Azure Container Instances** (**ACI**): Excellent for containers and applications that are less complex

- **Azure Service Fabric** (**ASF**): More complicated, but good for containers that need to maintain state

- **Web Apps for Containers**: Typically used for less complex container-based applications

- **Azure Spring Apps**: Meant for running Spring Boot and Spring Cloud applications in Azure

- **Red Hat OpenShift**: A specialized implementation of Kubernetes that has some Enterprise features

Now, let's take a deeper look at microservices. One of the leading proponents of microservices is Martin Fowler. He describes them in his *Microservices Guide* (`https://martinfowler.com/microservices/`) as follows:

> *"The microservice architectural style is an approach to developing a single application as a suite of small services, each running in its own process and communicating with lightweight mechanisms, often an HTTP resource API. These services are built around business capabilities and are independently deployable by fully automated deployment machinery. There is a bare minimum of centralized management of these services, which may be written in different programming languages and use different data storage technologies."*

From his statement, you get an idea of how microservices differ from traditional monolithic architectures and how they can provide advantages for modern applications. Let's take a look at some of the key advantages:

- **Application scalability**: Each microservice component can scale independently from the others. This allows for better resource utilization, thus allowing better application scalability.

- **Application agility**: Since each microservice is independent, we can introduce new features and functionality to the application without affecting the application as a whole. This means entire development teams can implement a feature update without incurring an outage or disruption.

- **Resilience**: Microservices are built to withstand failures. If one component fails, the rest of the system can continue to operate normally, and the failed component can be replaced or repaired without it affecting the rest of the system.

- **Technology-agnostic**: Microservices architecture permits the use of different programming languages and technologies for various application components, facilitating the use of the optimal tool for each task.

- **Better fit for modern DevOps**: Teams can deploy individual components separately thanks to the microservices design, which is in alignment with DevOps and continuous delivery techniques. As a result, releases are made more frequently and occur more quickly.

- **Increased modularity**: Microservices design enables a more modular approach to application development, which ultimately results in an application that is less difficult to administer and maintain over time.

As monolithic architectures grew over time, they evolved into N-Tier architectures. These evolved from a monolith into 3-Tier, and finally N-Tier for more distributed deployments. With this approach, the different application functionality was put on separate servers. They are typically separated as follows:

- **Presentation tier**: This tier addresses the UI and API components to allow different user interfaces and application access

- **Application tier**: This is where the application logic and state management reside

- **Database Tier**: As the name implies, this is where the storage and database reside

Even though the functionality was separated on different servers, this was and is still considered a monolithic architecture. When we endeavor to migrate from a monolithic or an N-Tier architecture to a microservices-based one, we need to *decompose* the modules within the monolith to independent microservices. Decompose here means that we are taking the complex application and breaking it down into smaller more manageable applications, or services.

The following figure illustrates what the transformation looks like at a high level:

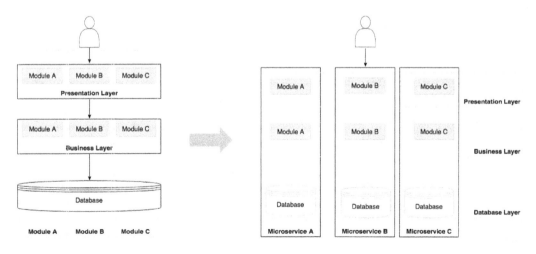

Figure 10.2 – The monolith to microservices transformation

Now that we've dealt with microservices, let's look at the Saga pattern.

What is the Saga pattern?

If you ever did application development on a monolithic or an N-Tier system, you are probably familiar with the concept of a transaction. This is where the application wants to do a **Create, Read, Update, or Delete (CRUD)** type transaction on a database or multiple databases. It needs to do this consistently across all participating databases. A transaction manager was typically used. It ensured that the transaction as a whole either happened or didn't happen. The term to refer to this is an ACID transaction. Let's look at what the ACID acronym stands for:

- **Atomicity**: Either the entire transaction happens or does not. There is no in-between state.

- **Consistency**: The data being affected by the transaction is in a consistent state both when the transaction starts and when it ends.

- **Isolation**: When a transaction is in progress, the state of the transaction is invisible to other transactions that might be running.

- **Durability**: After the transaction alters the data, the data cannot revert to its previous state. It remains as-is.

On mainframes, the solution that provided this functionality was **Customer Information Control System (CICS)**. Additionally, transaction managers were integrated into database platforms and took part in a **distributed transaction**, a much more complicated process. This is a transaction that affects multiple, potentially disparate, databases in the context of a transaction. This means that everything we talked about for ACID transactions has to happen on multiple databases.

When we move to a microservices-based architecture, we need to rely on services for this similar functionality. The pattern that allows us to achieve this and other scenarios as well is the **Saga pattern**.

The Saga pattern is a method for managing transactions involving multiple services. Different services may need to collaborate in a microservices architecture to fulfill a larger transaction. Consider a consumer who wishes to purchase a product from an online store. This transaction would entail multiple services, including the service for cataloging products, the payment service, and the shipping service.

Each service involved in the transaction has a local transaction in a Saga pattern. If any of the local transactions fail, a compensating action is initiated to undo the modifications made by the preceding local transactions. This enables the transaction as a whole to succeed or fail, ensuring that the system remains consistent.

Continuing with the example of the online store, suppose the payment service is unable to conclude its local transaction. The compensating action would then start to reverse any alterations made by the previous local transactions, such as canceling the order or refunding the customer's money. This ensures that the system is consistent and that the consumer is not left with an incomplete transaction or charged for an incomplete purchase.

So, you will need to be familiar with the Saga pattern in your transformation journey. Another pattern you should be familiar with is the Strangler Fig pattern. Let's check it out.

What is the Strangler Fig pattern?

To understand what the Strangler Fig pattern is, it is probably best to describe where it derived its name from.

In nature, a fig tree often begins to grow on another tree (the host tree) and finally wraps its roots around it, slowly strangling and engulfing the host tree's growth. This phenomenon is known as the **Strangler Fig pattern**. The host tree is eventually killed and replaced by a new fig tree as a result of the strangler fig's roots absorbing nutrients and water from it. This pattern is an illustration of the natural process of resource rivalry, in which one species subjugates another by using its resources.

In software development, the Strangler Fig pattern is used as a metaphor to describe how a new software system replaces an older one by gradually absorbing its resources and functions without substantially altering the previous system.

With this pattern, it is implied that certain services will be migrated first and others after. To have a holistic application that is technically in two worlds (that is, one legacy and one modern), we need a proxy or service facade layer that can redirect application traffic to the appropriate service, be it on the mainframe or in Azure. The following figure depicts how this proxy would look when implemented:

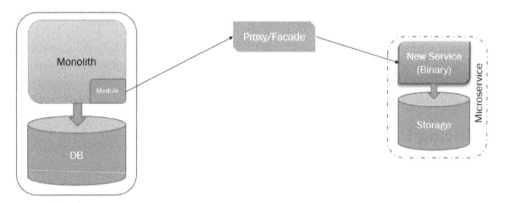

Figure 10.3 – The Strangler Fig pattern

This method helps spread out the development work over time and reduces migration risk. You may add functionality to the new system at any rate you like while ensuring the legacy application continues to work because the façade safely directs users to the appropriate application. The legacy system gradually becomes *strangled* and is no longer required over time as features are transferred to the new system. After finishing this procedure, the legacy system can be safely retired.

In the previous topics, we looked at two patterns that you will find useful for transforming a monolith into microservices. Patterns are essentially approaches to how to develop your application. There are other patterns you will need to be familiar with as well. Here is a list with a brief description of each:

- **Service Registry pattern**: A centralized repository that may be used to find microservices by their names is provided by the Service Registry pattern. It is a design for the architecture of microservices that enables services to discover and communicate with other microservices.

- **Circuit Breaker pattern**: As the name implies, the Circuit Breaker pattern interrupts the circuit to stop a cascading failure from occurring. This allows applications to continue operating even if one or more services fail. In a microservice architecture, it is employed to handle errors should they occur.

- **Command Query Responsibility Segregation (CQRS) pattern**: In the architecture of some applications, the processes of reading data from the database and writing data to the database can be very distinct from one another. The CQRS pattern is a design pattern that divides commands (write actions) and queries (read operations) into distinct models, each of which has a database.

- **Bulkhead pattern**: This design pattern is a method of compartmentalizing a system so that the failure of a single component won't bring everything down. To prevent the failure of an entire system due to the failure of a single microservice, the Bulkhead pattern can be implemented in a microservice architecture.

There are other patterns that you will surely come across. These are just a few of the more common ones. Now, let's look at an actual toolkit that can help you implement these patterns.

In the course of migrating your legacy workloads to Azure, you will need to have your new Azure deployed workloads communicate with the workloads still on the mainframe and vice versa. Typically, this communication is done via service calls via an API. To do this communication seamlessly, you will need a tool such as Dapr.

Introducing Dapr

So, what is Dapr? **Dapr** was created by Microsoft and then released to the open source community. It is a framework for addressing issues with service-to-service communication, state management, pub-sub messaging, and actor programming models, which are typical in distributed systems. It is relevant in several scenarios, but for our focus, it can also be leveraged for legacy transformation to modern architectures, especially for implementing the Strangler Fig pattern.

It offers SDKs to make it easier to develop software in several languages, including Java, and supports a wide range of programming languages, such as Java, Python, NET, Node, Go, Rust, JavaScript, and others.

Dapr is helpful for application modernization since it frees developers from caring about infrastructure and platform-specific issues so that they can concentrate on building pure business logic. It provides a collection of standardized APIs that hide the infrastructure's underpinnings, making it simple to install and manage applications across various contexts.

The following are some advantages of utilizing Dapr for application modernization:

- **Increased agility**: Dapr gives developers the ability to create applications in a language-independent manner, giving them the freedom to select the right tool for the job. This has the effect of accelerating development and enhancing agility.

- **Easier development**: Development is made easier thanks to Dapr's set of pre-made components and APIs, which make it simple to implement popular patterns for microservices-based applications.

- **Better scalability**: Dapr offers a distributed runtime that is scalable and can manage high-traffic applications, making it the perfect choice for contemporary applications that must expand rapidly and effectively.

- **Flexibility**: Dapr was created to be cloud-agnostic, which enables it to function on any on-premises or public cloud architecture. With this flexibility, you can update your applications without being restricted to a single platform.

In conclusion, Dapr is an effective tool for updating your applications. It makes it simple to design, deploy, and maintain distributed applications by providing a collection of pre-built components, standardized APIs, and a runtime independent of any particular cloud.

Other essential microservices components

We cannot have a chapter on modernizing a monolith without a deeper dive into three substantial topics that you will need to consider in your journey. Let's look at them.

Azure API Management gateways

An API gateway is an essential part of a design that uses microservices. It is the entry point that lies between clients (such as web browsers and mobile applications) and the microservices that come together to form an application. It is the role of the API gateway to take in requests from users and then forward those requests to the relevant microservices. In essence, it abstracts the applications from the application users. Requests that come in are received by an API gateway, and then that gateway determines which microservice is best suited to handle them depending on the URI or path of the request.

The following is a list of the primary functions of an API gateway:

- **Load balancing**: Load balancing is the process by which incoming requests may be distributed evenly among many instances of a microservice via an API gateway. This helps enhance both performance and availability.

- **Authentication and authorization**: The API gateway can authenticate clients and authorize access to microservices depending on the credentials and permissions that are associated with the client.

- **API versioning**: An API gateway may handle different versions of the same API and divert requests to the appropriate version of the API based on the client's request. This feature is referred to as **API versioning**.

- **Traffic management**: The API gateway can control the traffic between clients and microservices. This management includes throttling and rate-limiting to avoid the microservices from becoming overloaded.

- **Caching**: To increase response time and lessen the stress placed on the microservices, data that is often requested might be cached by an API gateway.

In a microservices world, an API gateway is essential. Services are often deployed independently in a microservices architecture, and they communicate with one another using APIs. This technique permits flexibility, scalability, and maintainability in its implementation. On the other hand, administrating and protecting each of these unique services may become difficult to do, particularly as the number of services expands. An API gateway manages the traffic and ensures the security of the communication that takes place between clients and services. This helps simplify the complexity that is thereby acting as a single point of access to the application for users. It frees developers to concentrate on creating and delivering particular services without requiring them to think about how customers will access and use those services. This makes it possible for developers to work more efficiently.

Event-driven architecture versus traditional batch processing

Legacy applications, especially mainframe legacy applications, rely heavily on traditional batch processing. **Batch processing** is the method of data processing in which information is gathered and processed in bulk at predetermined intervals. Typically, there is a series of *batch* jobs that run at night, often called nightly batch jobs. These batch jobs process the online transaction that occurred during the day.

With the advent of microservices, applications can now react in real time to events or messages thanks to **event-driven architecture**, yet another pattern. By eliminating the need to wait for the completion of a batch operation, event-driven architectures can take the role of batch processing. Faster reaction times, processing data in real time, and increased scalability are just a few of the advantages this strategy may provide.

Applications with an event-driven architecture are made to react to certain events as they occur. As soon as an event occurs, the application reacts to it by doing something such as updating a database or starting a new process.

This method can be especially helpful for data-intensive or time-sensitive applications such as online marketplaces and stock trading platforms. An event-driven architecture processes data as it comes, allowing faster insights and immediate actions than are possible with batch processing.

In your modernizing journey, you should consider replacing some, or possibly all, of your batch processing because, in comparison, event-driven systems are more adaptable and timelier in their data processing. They improved speed, have less latency, and yield more timely insights.

The rise of Apache Kafka

When putting together this chapter, I was compelled to put in a brief topic on Kafka, purely for its rise in popularity in the mainframe modernization space, especially in the financial services industry. It is fitting to put this topic right after event-driven architecture because, at its heart, that is exactly what Kafka is.

Essentially, **Apache Kafka** is a distributed streaming platform that was developed to manage real-time data feeds and process massive amounts of data in a manner that is scalable and error-tolerant. As a messaging system that allows services to connect with one another and enables data streaming, Kafka is a component that is widely used in microservices architecture. Kafka serves as a communication layer between services, making it possible for services to trade data and events with one another.

A concept known as **publish-subscribe** serves as the foundation for Kafka. In this model, producers send messages to a topic, and consumers subscribe to the topics to receive those messages. This enables services to interact asynchronously, and it also decouples the producer from the consumer, making it possible for services to function independently of one another.

In addition to its messaging features, Kafka is also utilized for real-time analytics, data processing, and data integration, because it supports numerous programming languages and platforms, as well as data replication, fault tolerance, and scalability.

In summary, Kafka offers a dependable and scalable method for microservices to interact with one another and share data. As a result, it is an essential component of many contemporary systems that make use of microservices.

Now, let's take a step back and consider what we have just explored and see how it fits into the large view of modernizing a monolith.

In modernization, one size does not fit all

Everything we have talked about so far has been focused on re-architecting your monolithic workloads. You might think that this implies that you are rewriting most, if not all, of the workloads, but it does not have to always be that way. It's time for a quick recap to make sure we are clear on semantics. Here are a couple of terms to review:

- **Rehost**: Typically, this refers to taking the existing code base from the monolith and recompiling it for the x86 platform so that it can run in Azure

- **Refactor**: This is very similar to rehost but implies that the language has changed, usually from COBOL (or some other legacy language) to Java or C#

Remember earlier in this chapter in the *Understanding how we got here – the evolution of application development* section, how we talked about how, as monoliths evolved, they started employing modular programming into their architectures? This is good news if we want to reuse that code because often, the modular approach lends itself nicely to a service-oriented and microservices architecture. What this means is that we can often leverage that code base in our re-architecting transformation of the monolith. This is something to be aware of in your discovery process. Keep in mind that there are solutions in the marketplace that allow COBOL to be compiled into .NET and deployed in containers. This means that you can have microservices written in COBOL. That might seem strange at first, but the solutions that are in the marketplace are very mature.

The following figure shows the migration paths you can take on your transformation journey:

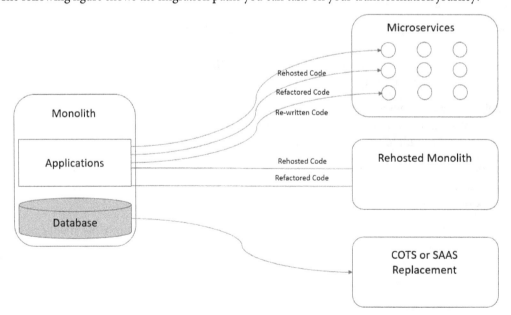

Figure 10.4 – The migration paths available for modernizing the monolith

The point here is that microservices are not exclusive to only rewritten code. You can often leverage your existing code base and migrate it to a microservices architecture. This could save a substantial amount of time in your transformation from the monolith to microservices.

Best practices for modernizing the monolith

The transition from a monolithic design to one that is built on microservices can be a difficult process that calls for careful preparation and execution. The following are some examples of excellent practices to think about:

- **Know your monolith**: Before implementing any sort of modification, you should make the effort to completely comprehend your existing monolith. Determine the various components, the dependencies between them, and how they interact with one another. This will assist you in determining which components of the system can be broken out into independent services (microservices).

- **Know your microservices' capability**: Determine the limits of your microservices based on the business capabilities of your application. To do this, you must first identify the business capabilities of your application. These should be logical collections of related functionalities that can be deployed, scaled, and maintained separately from one another.

- **Start small**: Begin by partitioning your monolith into microservices by breaking off discrete, well-defined components. Before going on to larger components, you will have the opportunity to evaluate the process and improve your strategy.

- **Utilize API gateways**: If you want to create a unified interface for your microservices, you can use API gateways. The development process may be streamlined as a result, and the management of your microservices may become less complicated.

- **Establish a culture of DevOps**: A DevOps culture emphasizes collaboration, communication, and automation between the teams responsible for software development and system administration. This is a crucial component for effectively managing an architecture based on microservices.

- **Utilize containerization technology**: Containers should be utilized because they offer a method that is both lightweight and portable to bundle and deliver microservices. When thinking about how to manage your microservices, you should think about adopting a containerization technology such as Docker or Kubernetes.

- **Maintain data integrity**: One of the problems of developing an architecture based on microservices is maintaining data integrity across several different services. Use techniques such as event-driven architectures, message queues, and transactional databases to assure data consistency.

You will be able to successfully upgrade your monolithic workloads into ones that are built on microservices if you keep these best practices in mind.

Summary

In this chapter, we explored some of the most important topics you will need to consider when modernizing your monolith. Much of the microservices world is still emerging, so you will need to do your research to decide which of the topics we covered are relevant to your journey. Keep in mind, as we stated earlier in this chapter, that there is no one size fits all approach here. There are multiple paths to consider based on your vision of where you want to go and the workloads you are considering for modernization.

In the next chapter, we will wrap things up with a look at trends and emerging technologies that you need to be aware of to have a successful transformation journey.

11

Integration – Playing Friendly in a Cloud-Native World

Let's start by defining our view on what playing friendly in a cloud-native world means. In a cloud-native world, applications should have the following characteristics related to integration:

- Published interfaces for other systems to connect

- **Extensible connectors** to other systems

- The ability to scale up and down as needed

- Offer entry, standard, and premium services to meet customer needs

- Leverage **Platform as a Service (PaaS)** capabilities

- Isolation of services – limited or no cross-services dependency

- Integration with cloud monitoring services, such as Azure Monitor

The most important features needed for integration to play friendly in a cloud-native world are the published and extensible interfaces (items one and two in the preceding list). These allow your application to integrate with other systems in a prescribed and predictable manner for both incoming and outgoing information and data. **Extensibility** is important to allow integration into new applications and data sources. Extensibility is often accomplished by having a "driver" model, where a new driver is created for new applications and data sources as needed.

Implementing the other items listed previously typically involves using existing Azure services once the interface model and extensibility model have been created. The following diagram shows the various types of extensible services you might use in an application modernized to Azure:

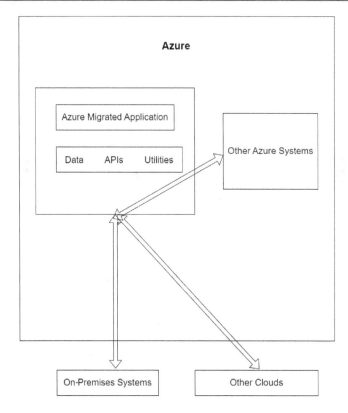

Figure 11.1 – Azure integration options overview

For the remainder of this chapter, we will look at various categories of integration your application might need, including data, other applications, on-premises integration, and third-party integration.

This chapter covers the following topics:

- The role of integration in a cloud-native world
- The different types of integration
- How integration fits into the bigger picture

Now that we've provided an overview of the various integration services an application typically uses, next, we will look at the role integration plays in a cloud-native world.

Technical requirements

The technical requirements for integration will depend on both the functionality of the application and the architecture in which it is deployed.

The role of integration in a cloud-native world

Most new applications developed for Azure use Azure-native tools such as **Azure Kubernetes Service**, **Service Bus**, **Service Fabric**, and **Azure Functions**, and low-code solutions such as **Logic Apps** and **Power Apps**. However, many existing legacy applications in a typical customer's estate may also either be hosted in Azure or integrated with Azure. This is where Azure integration tools come into play. Both Microsoft and third parties provide tools for integration. These tools roughly fall into four categories.

Data integration

Data integration is probably the most important thing we need to interoperate with Azure-native systems. This can take several forms. We will cover the most common we typically see. These include the following:

- **Replication**: This could include **snapshot replication** (entire database), **delta replication** (only changes since the last sync point), and **transactional replication** (changes when they are committed in the source database). **Azure SQL Database** and **SQL Managed Instance** provide these capabilities, as do third-party tools.

- **Change data capture** (**CDC**): This is a specific type of real-time replication of data. This is provided by SQL Server and third-party tools that capture changes in the databases' transaction logs.

- **ETL/ELT**: Tools to export, transform, and load data are available with PaaS services such as **Azure Data Factory** (**ADF**), **SQL Server**, and third-party tools.

- **Direct data access**: This is performed using data providers such as OLE DB and ODBC. This can be used with a variety of data systems on Azure, including transactional, reporting, and data warehouse solutions.

- **Distributed Relational Database Architecture** (**DRDA**): DRDA is a standard from the Open Group for bi-directional data interchange that is available on mainframes for Db2 and can be implemented on Azure using the DRDA feature of **Host Integration Server** (**HIS**).

- **Streaming data**: Solutions such as **Azure Stream Analytics** (**ASA**) or **Kafka**.

- **Queuing**: Services such as **Service Bus** and third-party querying tools.

- **Low-code solutions**: An example of this is **Logic Apps**.

Having all these methods available means that Azure probably has a method that is compatible with both legacy systems and Azure-native systems. Our recommendation is to use a PaaS solution such as ADF, Service Bus, ASA, or Logic Apps as these solutions require less overall support, and leverage Azure capabilities.

The following diagram shows the various data services in Azure, and provides a list of potential sources of data for these services:

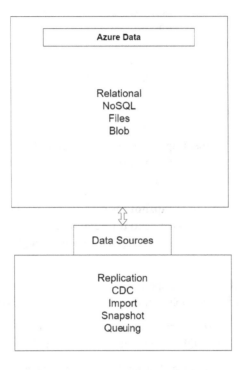

Figure 11.2 – Data integration in Azure

Now that we've looked at data integration, we will explore options for application integration.

Application integration

For this chapter, application integration is defined as an **application programming interface** (**API**). This can be achieved with direct API calls from one application to another, or via some type of **API gateway** or **API harmonization layer**. The preferred method to be cloud-friendly is to abstract the applications from direct API calls and use an API gate or API harmonizing layer.

Direct application-to-application API calls are not the preferred method to be cloud-native friendly for the following reasons:

- Direct API calls require dependencies between applications that can limit their ability to be deployed and scaled in a cloud-native environment.

- Direct API calls are not extensible in terms of including additional applications and services. Each application requires a unique API.

- Not all applications have a published API.

- It is not easy to change from one API provider to another. Additional programming is usually required.

Next, we'll look at the API abstraction concepts of API gateways and API harmonization. To start this, we'll provide an overview of the two approaches.

API gateways allow interfaces for a specific platform to be published in a manner that abstracts the underlying API provider (server) from the consumer (client). Azure API Gateway in Azure and z/OS Connect on the mainframe are both examples of API gateways.

API harmonization layers are systems that not only interface with published APIs, but also take input/output and make any translations needed to complete the application-to-application integration.

Here are some examples of transformation/harmonization that could take place:

- Mapping user IDs between systems.

- Machine-level code translations, such as **American Standard Code for Information Interchange (ASCII)** and **Extended Binary Coded Decimal Interchange Code (EBCDIC)**. This is how data is typically coded in eight-bit format for use in computers.

- Mapping data types.

You can think of API harmonization as taking an extra step beyond what an API gateway does. An API harmonization layer may be either a provider or a subscriber to an API gateway. API harmonization layers are often unique to specific customers and may be built on Azure using tools such as **Distributed Application Runtime (Dapr)**.

The advantages of taking the API gateway and/or the API harmonization approach include the following:

- Published interfaces

- Extensible across both platforms and applications

- Publisher/subscriber model-compatible

- Allows transformations, if needed

- Allows user ID and security mapping

- Can be implemented using Azure PaaS services

With all of these advantages, the API gateway/harmonization model is the preferred method for playing friendly with API integration in a cloud-native environment using Azure.

Monitoring and management integration

Another way for legacy applications to play friendly in an Azure-native environment is to allow for application and system monitoring and management across legacy and Azure environments, and even to allow that capability across other clouds, such as **AWS** and **GCP**. This approach recognizes that Azure does not live in an isolated environment and needs an integrated approach for monitoring and management across both a hybrid cloud and a multi-cloud estate for a customer.

Azure has a service called **Azure Arc** for this type of hybrid/multi-cloud management approach. Azure Arc already supports several different systems across different platforms. Additionally, Azure Arc has new capabilities added when demand is there. Azure Arc represents the concept of system management playing friendly in a cloud-native environment.

Application operations integration

The final type of integration we will look at is how to play friendly with legacy operations in a cloud-native environment. What we mean by integration with legacy operations includes things such as the following:

- Scheduling
- Backup recovery
- DevOps integration

For scheduling, the cloud-native approach involves viewing scheduling as more of a set of tasks that are run based on a series of events. The way to play friendly with scheduling in a cloud-native environment would be to use tools such as **Event Hubs** and Kafka in Azure and to map events triggered in these tools to specific jobs or applications that need to run. Eventing tools could also be combined with **Azure Scheduler** to allocate resources or pools of resources needed to run the applications or jobs that are triggered.

With backup and recovery, the way to play friendly with a cloud-native environment in Azure is to integrate with **Azure Blob storage**. This has the following advantages:

- Azure Blob storage is typically less expensive
- Azure Blob storage has hot, warm, and cold options, allowing data to move to less expensive versions of Blob storage over time based on the policy
- Azure Blob storage supports several interfaces, such as using **Representational State Transfer (REST)** APIs and **Network File System (NFS)**
- Azure Blob storage has options for both redundant storage within a region and cross-region redundant storage for disaster recovery scenarios

Finally, **Azure DevOps**, **DevSecOps**, and **GitHub** allow you to integrate modern DevOps and storage in both cloud-native and legacy development environments. Many legacy development environments already provide integration capabilities with Azure DevOps and GitHub. This is the recommended way to play friendly in a cloud-native way with legacy development environments. Azure DevSecOps is a specific way to use Azure DevOps for security policies, changes, and change tracking.

With that, we have finished looking at how various forms of integration take place within Azure, both with a hybrid implementation and a multi-cloud customer estate using a cloud native-friendly approach. Next, we will look at the various types of integration based on scenarios.

The different types of integration

What we mean by types of integration in this section is the types of systems that would possibly need to integrate with a newly transformed system from a legacy platform to Azure. These include the following:

- Other legacy platforms in a customer's estate, such as mainframes, **UNIX**, **Linux**, and **Windows**

- Other systems running Azure

- Systems running in other clouds

- Integration with third-party enterprise systems such as **Enterprise Resource Planning (ERP)**

So far, we have looked at integration from the point of view of data and applications. Next, we will look at integration between Azure and on-premises systems.

Integration with the existing legacy estate

Let's start with mainframes. Just like with any other platform, mainframes have data, applications, and utilities. Let's look at the options for dealing with each of these.

Data

Mainframes have several types of data sources that can integrate with Azure:

- **Relational databases**: Primarily, this is Db2. Db2 offers several methods for integration, including **replication**, **CDC**, **import/export utilities**, and **DirectQuery**. Microsoft, with systems such as HIS, and third parties offer systems to enable this functionality.

- **Virtual Storage Access Method (VSAM)**: You can think of VSAM as a special type of flat file that allows indexed, relative, and sequential access. Usually, VSAM data is either imported or exported to Azure. There are additional tools that allow replication.

- **Other flat files**: Mainframes can also have several sequential flat files. Integration with these is usually done with import/export utilities.

- **Information Management System (IMS) DB**: This is a hierarchical database on the mainframe. This usually requires either some type of third-party import/export utility or using import/export utilities.

- **Third-party databases**: There are several third-party databases, such as **Datacom, Integrated Database Management System (IDMS)**, and **Adabas,** that run on the mainframe. Utilities from the **independent software vendors (ISVs)** that market these databases are usually required.

In many of the preceding data sources, the option for import/export exists. This allows data to be exported from the mainframe. Here, some type of filesystem transfer will be required. For this, we have the following options:

- **File Transfer Protocol (FTP)** and **Secure File Transfer Protocol (SFTP)**

- NFS

- **Server Message Block (SMB)**

The best method for you will depend on what is available on the mainframe. Keep in mind that you may need to translate from EBCDIC to ASCII to use this data.

UNIX and Linux systems, in addition to mainframes, also run relational databases. These include the following databases:

- **SQL Server**

- **Oracle**

- **Sybase**

- **Db2 LUW**

- **Sybase**

- **PostgreSQL**

- **MySQL**

- **MariaDB**

- **Informi**

All of these offer varying degrees of **direct data access**, replication, **CDC**, and import/export. UNIX, Linux, and Windows systems also have flat files. FTP, NFS, and SMB are the normal protocols that are used for file transfer.

In all these cases, ADF can be used to orchestrate the data exchange between legacy systems and Azure.

Finally, latency (the elapsed time between input/output operations), or controlling the effect of latency, is one area that needs particular attention when dealing with data on non-Azure data sources. Of particular concern is limiting redundant or chatty (asking for several instead of one) data requests.

The following diagram shows the different types of data storage in Azure and the sources that use these types of data:

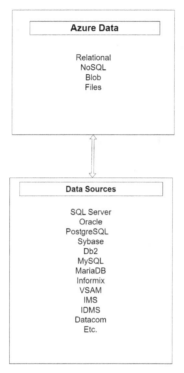

Figure 11.3 – Data sources and types

Let's move on to the applications.

Applications

Integration with legacy systems may also occur using the API. Some common interfaces are as follows:

- **REST calls**: Available with most legacy systems that have recent updates
- **Customer Information Control System (CICS) calls**: Normally done with Microsoft or third-party tools
- **Information Management System Transaction Monitor (IMS TM) calls**: Normally done with Microsoft or third-party tools

- **Third-party application servers**: Usually require third-party tools
- **Service brokers**: Require third-party tools

Let's move on to the utilities.

Utilities

This includes services such as **queuing** (usually with **MQSeries**), or other utilities such as backup/recovery, scheduling, monitoring, and management. These systems typically map to Azure services as follows:

- **Queuing**: Azure Service Bus
- **Events**: Azure Event Hubs or Kafka
- **Backup/Recovery**: With Azure Blob Storage
- **Scheduling**: Azure Event Hubs, third parties, and Azure Scheduler
- **Printing**: `.pdf` software and Teams/Sharepoint
- **Hierarchical storage management** (**HSM**): An Azure Blob storage aging policy

Now that we've seen integration with the existing legacy estate, let's consider integration with other Azure-based systems.

Integration with other Azure-based systems

Integration with other systems in Azure falls into two general categories. The first deals with systems in Azure that use native Azure tools such as **SQL DB** and **SQL MI**. The second deals with using **Infrastructure as a Service** (**IaaS**) services on Azure.

IaaS systems

In addition to PaaS services, you might also have several systems in Azure implemented with **virtual machines** (**VMs**). These are often referred to as IaaS systems.

This form of integration is essentially the same as dealing with non-mainframe systems on-premises, with the exception being that latency is no longer a big issue.

PaaS systems

For PaaS systems for data such as SQL DB, SQL MI, and Cosmos DB for NoSQL, the available options, in addition to the ones available for IaaS systems, include using other PaaS systems in Azure, such as Azure Service Bus and Azure Event Hubs, to allow for easier integration between applications.

Integration with other clouds

Other clouds include AWS, GCP, **z/Cloud**, and additional cloud platforms. From an integration approach perspective, systems running on other clouds are very similar, if not identical to those running other systems on-premises. Here are some things that need to be accounted for:

- Protocols for data transfer, such as FTP, NSF, or SMB

- Third-party utilities to extract data

- Allowances for latency

Now, let's check out integration with third-party systems.

Integration with third-party systems

Some third-party enterprise applications may require integration with an Azure-based solution. The third-party application might run in any of the options already discussed in this section. In particular, high-volume applications such as ERP, healthcare systems, and large banking systems have specific required interfaces for both data input and inquiry.

The guidance for integration with these systems is similar to our guidance for non-PaaS relational databases, but here, we should pay particular attention to latency and the effect data queries might have on overall performance.

Now that we have looked at the various types of integration an Azure-based application might encounter, next, we'll look at the role integration plays in the overall migration process and subsequent operations.

How integration fits into the bigger picture

Since most systems do not exist in a vacuum, planning for integration with your new system in Azure is extremely important. The success of integrating with other applications and data sources can make or break the modernization project when moving a legacy system to Azure. For this reason, integration needs to be a part of a modernization project from the initial design, through development and deployment and in operation after deployment.

Fortunately, Azure provides PaaS services to enable integration and the ability to host third-party integration solutions in both VMs and containers. The final two sections of this chapter will cover the things you should do and what to avoid when creating integration and playing friendly with systems in Azure.

Strategies for playing friendly

Strategies for playing friendly revolve around understanding some of the basic concepts of Azure native solutions. Let's take a look:

- **Support scale-out capabilities**: In other words, allow additional storage and compute for your solution without changing the system's architecture

- **Support scalability on demand**: Allow for **dynamic scale-up (bursting)** and scale-down based on application load and demand

- **Support location independence**: Azure-friendly systems can be deployed in multiple availability zones within a region, as well as across multiple regions

- **Use recovery points**: Implement your integration solution with recovery points for easier restart in case of failure

- **Support failover and recovery**: Incorporate **Azure system recovery (ASR)** and regional Azure storage options as part of your integration design

- **Externalize state when possible**: Use capabilities such as **Azure Redis Cache** to store state

- **Use Azure PaaS tools whenever possible**: Using PaaS integration tools such as ADF, Azure Event Hubs, and Azure PaaS data solutions to enable integration will automatically include Azure-friendly capabilities in your solution

Next, we will look at what to avoid.

How to avoid not playing friendly

Just as there are things to do for integration to play friendly in an Azure-native ecosystem, there are also things not to do. The following is a list of common things to avoid:

- **Avoid chatty interfaces**: When integrating from on-premises and other clouds to Azure, each round trip incurs a latency penalty. Using a more *chunky* or batch data flow will improve performance.

- **Avoid transaction interfaces**: Any type of transaction that crosses location boundaries is more problematic and prone to failures. Use compensating actions to enforce consistently instead.

- **Avoid recovery from the starting point**: Implement recovery/restart points in your interfaces. Without this, systems needed to be completely rerun.

- **Avoid scale-up solutions**: Cloud systems are meant to scale out and use dynamic scaling, even for integration work. Scale-up does not take advantage of compute options, such as Kubernetes clusters, Spark clusters, and Azure Service Fabric.

- **Avoid interfaces to application-specific APIs**: Instead, use published interfaces. Otherwise, you will limit the ability to integrate with other systems.

- **Avoid inter-application dependencies**: Instead, use published interfaces based on the work to be done to allow for options to change out the applications on the other side of the interface. Otherwise, changes to one application may have unexpected consequences for another application.

Now that we have seen what to do and what to avoid with integration in an Azure-friendly environment, let's summarize this chapter.

Summary

As we've discussed in this chapter, integration plays a key role in the success of most systems that move from their legacy environment to Azure. To play friendly with integration in Azure, we need to know and follow these rules. Using the information provided, you can now implement flexible and scalable interfaces to your Azure systems that run well in a heterogeneous estate. At this point, you understand where the systems exist in the organization's estate, such as on-premises, other clouds, or within Azure. You also know how to use Azure-friendly integration tools. You should use Azure PaaS services where possible; otherwise, use tools that leverage cloud-native implementation techniques such as auto-scaling and clustering. Finally, you know that you should avoid non-cloud-friendly integration tools when possible.

Now, let's move on to the final chapter, where we will discuss the trends in legacy modernization to Azure, as well as provide some thoughts on future capabilities.

12

Planning for the Future – Trends to Be Aware of

So, here we are: the final chapter. Hopefully, this book has given you some insight into legacy modernization and the challenges involved when migrating from a legacy system to Azure. This book intended to give you an overview of the types of systems you might encounter and the special considerations you will need to be aware of when migrating from them. We also wanted to provide you with some guidance when it comes to the Azure platform, with the intent of making you aware of not only the **Infrastructure as a Service (IaaS)** services available but also the cloud-native services available for more modern application deployment.

But, if you have been in the technology field for any amount of time, you will know that the target is always moving. Sometimes, it moves very fast – as it is today with the advent of **artificial intelligence (AI)** and modern application architectures. And as they say in hockey, you don't skate toward the puck, you skate toward where the puck is going. This chapter will provide you with an overview of the trends you will need to be aware of on your transformation journey.

In this chapter, we will cover the following topics:

- Where we are today
- What to expect in the near term
- What to expect in the long term
- Areas of modernization innovation for Azure
- Where do Azure partners fit in?
- Where does Microsoft fit in?
- The advent of multi-cloud

Where we are today

Today, Azure can successfully run mainframe applications in the following ways:

- **Rehost**: Take COBOL/PL/I/Assembler on VMs and containers that provide source compatibility for these languages
- **Refactor**: Take legacy languages and refactor them into more modern languages such as C# or Java, then run the converted code in VMs, containers, or microservices
- **Rewrite**: Take legacy mainframe systems and rewrite them to be **Azure-native**

While these approaches can work for most legacy systems, there are several downsides. The following table compares the advantages and disadvantages of the currently available approaches:

Approach	Time to Complete	Cost	Risks	Comments
Rehost	9 to 24 months	Low to medium	Low	The fastest time to move to Azure, but it does not address technical debt.
Refactor	12 to 30 months	Medium to high	Medium	Gets rid of legacy languages. However, it takes longer to test, and refactored code may be difficult to maintain.
Rewrite	18 to 48 months	High	High	Requires additional design and time to develop, as well as additional validation and testing. May require personnel retraining.

Table 12.1 – Migration approach comparisons

In addition to migrating the source code, several other operational things need to be done. Currently, most operational functions require a mainframe-style tool for things such as scheduling, monitoring, and backing up/recovering, or require these tasks to be converted manually into Azure equivalents.

Both Microsoft and our partners are continuing to invest in making the move to Azure more streamlined and less expensive. The remainder of this chapter will look at things you can expect in both the near term and the long term. We will start by looking at things to expect in the near term.

What to expect in the near term

To be clear, let's define what **near term** means for the context of this book. We are in 2023 and for the most part, the near term can be considered to be the next 1 to 2 years. Most of what we have outlined and described in this book will still be very relevant. Of course, you would have to have been under a rock for the last year to not know about the advances in AI, especially with Microsoft and OpenAI. We will touch on this in the next section. This will no doubt affect code transformation and perhaps other areas as well.

The main vendor solutions are usually very mature and will be around for quite some time, but Azure, as with all cloud platforms, is evolving very fast. Now, let's take a look at what to expect in the long term.

What to expect in the long term

We believe that legacy modernization will become more about the best place to run an application or system rather than moving everything from one platform to another. This includes both the transactions between legacy platforms and clouds and between clouds. Most large customers have both a hybrid (cloud and on-premises) strategy and a multi-cloud strategy. Additionally, we see even greater interoperability between these environments. This can take on several different experiences, including the following:

- Management and monitoring from a single pane of glass (or dashboard). This includes both Azure and other hyperscale cloud providers. The best way to prepare for this type of journey to the future is with cloud-native deployments, such as those offered in Azure for both IaaS and **Platform as a Service (PaaS)** services.

- Incorporate no/low touch application generation for legacy code on Azure.

- We believe we will see more AI create code to either perform the functions of legacy systems or extend their functionality. The best way to prepare for this is to incorporate modern DevOps tools that use things such as GitHub OpenAI Copilot capabilities. The first step is to move your legacy code into modern DevOps.

- We believe data integration tools will continue to improve. This will allow data to exist in hybrid environments, which will ease the transition to Azure and between other hyperscale clouds. This will make latency and data conflict resolution more seamless.

- Finally, we will see legacy operating systems becoming more available as cloud services either in Azure or other hyperscale cloud providers. The decision then becomes choosing the right service to use for the application.

Now that we've talked about things we will see in the future, let's talk about how to set up systems in Azure to take advantage of both what's available today and what is coming in the near and long term.

Areas of modernization innovation for Azure

We've discussed several areas where we see innovation coming in the near and long term. Additionally, we think it is important to look at what Azure offers today, and how these may evolve in the future.

OpenAI and ChatGPT

As discussed in the previous section, Microsoft is committed to furthering **OpenAI** and **ChatGPT** in many different areas. One area, in particular, is modernizing legacy source code such as **COBOL**. Tools such as GitHub Copilot can incorporate OpenAI capabilities to streamline the modernization process.

Azure Arc

Azure Arc provides a method to manage resources, such as compute and data, across several different estates. This includes legacy environments and data, as well as other cloud environments. Azure Arc uses an agent that can be deployed in other environments that allow those environments to be managed in Arc. This kind of capability will make the transition from legacy systems to Azure more seamless since Arc can potentially be used for either environment. This means you will need to invest less time and effort in new tools and training for system management as part of the modernization process.

Azure Migrate

Azure Migrate provides tools to assist with modernizing Azure. These include things such as the following:

- Migration from on-premises data centers
- Enabling hybrid clouds as part of modernizing Azure
- Migration from other clouds
- Database migration to Azure

These tools and features receive updates when new services are available. Azure Migrate is a great place to start to see what is available.

Azure Data Factory

Azure Data Factory (**ADF**) provides a means to orchestrate data flow between other systems and Azure. Connectors are available from several legacy data sources, including mainframe, midrange, and Enterprise UNIX. New connectors are added as needed. ADF should be one of the first places you look when you need data integration services.

Azure Logic Apps and Azure Power Apps

Finally, another place to look for future enhancement to ease legacy modernization to Azure is to look to **Azure Logic Apps** and **Azure Power Apps**. These no-code/low-code solutions allow you to create Azure-native applications for functions currently performed by legacy systems without requiring a large staff of developers.

Logic Apps can create logical workflows based on the provided connectors. Power Apps can be used by developers, admins, and power users to create data drive applications.

Microsoft Fabric

Microsoft Fabric is a set of Azure services for data movement, AI, and analytics that's available as a single comprehensive service. This is a feature you should consider for the data that is part of a modernization to Azure. Microsoft Fabric provides the data and analytic services customers need to enhance their current data estate during the process of moving to Azure.

Now that we've reviewed some of the areas of innovation for legacy applications in Azure, let's review how partners fit in.

Where do Azure partners fit in?

Quite frankly, for most legacy transformation endeavors, you will need the services of an **Azure legacy transformation partner**. They play a crucial role in legacy modernization by leveraging their expertise and resources to help organizations migrate legacy workloads to Microsoft Azure.

Here's an overview of the role Microsoft Azure partners play in legacy transformation:

- **Assessment and planning**: Azure Legacy transformation partners work closely with organizations to understand their existing legacy environment, applications, and business requirements. They conduct a thorough assessment to identify the potential benefits and challenges of migrating legacy workloads to Azure. This includes analyzing code and application dependencies, data storage requirements, security considerations, and performance needs. Based on this assessment, the partners develop a comprehensive migration plan and strategy.

- **Migration execution**: Typically, once the assessment and planning phase is complete, partners assist in executing the migration plan. This involves various tasks, some of which we have covered earlier in this book, such as rehosting or replatforming the workloads to run on Azure VMs or containers, refactoring or modernizing applications using Azure services such as Azure Functions or Azure App Service, and rearchitecting or rebuilding applications using cloud-native technologies such as microservices or serverless architectures. There are lots of places where you might experience *bumps in the road* and this is where partners can leverage their expertise to ensure a smooth and efficient migration.

- **Data migration and integration**: Legacy transformation often involves migrating large volumes of data from legacy systems to Azure. Partners help with data migration strategies, including selecting the appropriate data transfer mechanisms, optimizing data formats, and ensuring data integrity and security during the migration process. They also assist in integrating the migrated data with other Azure services or applications to enable seamless operations and data access.

- **Application transformation and optimization**: Partners also help modernize legacy applications to take full advantage of the capabilities offered by Azure. This may involve refactoring monolithic applications into microservices architectures, containerizing applications using **Azure Kubernetes Service (AKS)**, or leveraging serverless computing options such as Azure Functions. They can help you identify opportunities for optimization, performance tuning, and cost savings in the new Azure environment.

- **Training and support**: Partners can provide training and support to ensure a smooth transition and help organizations effectively manage their modernized legacy workloads on Azure. One thing to keep in mind during a transformation/migration is that there is a *human factor* to consider as well. Users and administrators of the application will need to be trained and educated for the migration to be a success. Partners offer guidance on best practices for operating applications in the cloud, optimizing resource utilization, monitoring and troubleshooting, and security and compliance considerations.

- **Ongoing management and optimization**: After the legacy migration is complete, partners continue to provide support for ongoing management and optimization. This includes monitoring application performance, troubleshooting issues, applying updates and patches, optimizing resource utilization, and ensuring compliance with industry standards and regulations. They work closely with you to continuously improve and optimize performance, scalability, and cost efficiency.

So, as you can see, Microsoft Azure Legacy transformation partners play a critical role in guiding organizations through the complex process of legacy transformation by leveraging their expertise in Azure services, tools, and best practices to ensure a successful migration, application transformation, and ongoing management on the Azure platform.

Where does Microsoft fit in?

And let's not forget about Microsoft. Microsoft has teams in both engineering and field sales that focus on legacy and mainframe modernization. They can offer deep technical guidance on topics such as application migration and integration, data migration, and security. They publish guides and reference architectures to provide consolidated best practices and guidelines.

Here are some resources they have published, which can get you started:

- *Azure mainframe and midrange architecture design*: `https://learn.microsoft.com/en-us/azure/architecture/mainframe/mainframe-midrange-architecture`

- *Modernize mainframe and midrange data*: `https://learn.microsoft.com/en-us/azure/architecture/example-scenario/mainframe/modernize-mainframe-data-to-azure?culture=en-us&country=us`

- *Replicate and sync mainframe data in Azure*: `https://learn.microsoft.com/en-us/azure/architecture/reference-architectures/migration/sync-mainframe-data-with-azure?culture=en-us&country=us`

This guidance can be incredibly valuable for your legacy transformation journey.

More recently, Microsoft has provided **Azure landing zone accelerators for mainframe and midrange**. Essentially, these accelerators take the concept of the reference architecture to the next level by providing a Bicep template (which is more or less an abstraction of an ARM template) that implements the target architecture for you. One click of a button creates the reference architecture for you. The repository is maintained in GitHub and is constantly growing and evolving. When you are planning your transformation journey, you should look at the repository and see what architectures have been published that you can leverage. This greatly enhances and shortens the migration, saving you both time and money and providing a deployment with security and other best practices built in.

What is an **ARM template**, you ask? **ARM**, which stands for **Azure Resource Manager**, is a way to represent an implementable infrastructure within Azure. You might have heard of **Infrastructure as Code (IaC)**, and this is exactly what an ARM template is. You can leverage accelerators to create production environments, **User acceptance testing** (UAT), and development environments in Azure. You can also use them to set up **proofs of concept** (POCs) and pilot projects.

The advent of multicloud

So, fellow technology professionals, you have no doubt probably encountered the term **multicloud**. Let's take a look at how that might be leveraged for legacy modernization. To be clear, multicloud refers to the practice of employing multiple cloud service providers to host different parts, or the same part, of your infrastructure or applications. It involves distributing workloads across multiple cloud environments, such as Microsoft Azure, AWS, Google Cloud, or others, rather than relying on a single provider.

This is a growing trend in cloud computing. I have seen it used typically for customers who want to provide a failover region in another cloud provider. But there are other scenarios as well.

In the context of legacy modernization, multicloud can play a role in achieving a more flexible and resilient architecture. It allows you to leverage the strengths and capabilities of different cloud providers while reducing vendor lock-in and increasing overall reliability.

Azure Arc, which we talked about earlier in the *Areas of modernization innovation for Azure* section, can play a crucial role in your multicloud strategy.

When migrating legacy applications to the cloud, you can do things such as distribute workloads across multiple cloud providers based on factors such as cost, performance, geographic location, or specific services offered by each provider. This approach enables you to optimize resources and take advantage of the unique features and capabilities of each cloud platform.

In short, you can mitigate risks associated with relying on a single provider.

Multicloud provides you with options for flexibility, scalability, and resilience, allowing you to harness the benefits of different cloud providers while modernizing your legacy systems. Keep in mind, however, that the cost of this is additional cloud providers and complexity.

Summary

So, there you have it: a concise overview of the current trends in legacy transformation to be aware of. As you probably already know from being in the world of technology, things can change fast, especially now with the emergence of AI in aiding with legacy transformation. Where will we be in 2 years or 5 years? Who knows? It's very exciting. I hope you are looking forward to it as much as I am.

Hopefully, this book has provided you with some insight into how to plan for and execute your legacy transformation journey. Here's to a successful transformation. Happy trails!

Index

`Packtpub.com`

Subscribe to our online digital library for full access to over 7,000 books and videos, as well as industry leading tools to help you plan your personal development and advance your career. For more information, please visit our website.

Why subscribe?

- Spend less time learning and more time coding with practical eBooks and Videos from over 4,000 industry professionals

- Improve your learning with Skill Plans built especially for you

- Get a free eBook or video every month

- Fully searchable for easy access to vital information

- Copy and paste, print, and bookmark content

Did you know that Packt offers eBook versions of every book published, with PDF and ePub files available? You can upgrade to the eBook version at `Packtpub.com` and as a print book customer, you are entitled to a discount on the eBook copy. Get in touch with us at `customercare@packtpub.com` for more details.

At `www.packtpub.com`, you can also read a collection of free technical articles, sign up for a range of free newsletters, and receive exclusive discounts and offers on Packt books and eBooks.

Other Books You May Enjoy

If you enjoyed this book, you may be interested in these other books by Packt:

Microsoft Azure Fundamentals Certification and Beyond

Steve Miles

ISBN: 978-1-80107-330-1

- Explore cloud computing with Azure cloud
- Gain an understanding of the core Azure architectural components
- Acquire knowledge of core services and management tools on Azure
- Get up and running with security concepts, security operations, and protection from threats
- Focus on identity, governance, privacy, and compliance features
- Understand Azure cost management, SLAs, and service life cycles

Azure for Developers - Second Edition

Kamil Mrzygłód

ISBN: 978-1-80324-009-1

- Identify the Azure services that can help you get the results you need
- Implement PaaS components – Azure App Service, Azure SQL, Traffic Manager, CDN, Notification Hubs, and Azure Cognitive Search
- Work with serverless components
- Integrate applications with storage
- Put together messaging components (Event Hubs, Service Bus, and Azure Queue Storage)
- Use Application Insights to create complete monitoring solutions
- Secure solutions using Azure RBAC and manage identities
- Develop fast and scalable cloud applications

Packt is searching for authors like you

If you're interested in becoming an author for Packt, please visit `authors.packtpub.com` and apply today. We have worked with thousands of developers and tech professionals, just like you, to help them share their insight with the global tech community. You can make a general application, apply for a specific hot topic that we are recruiting an author for, or submit your own idea.

Share Your Thoughts

Now you've finished *Modernizing Legacy Applications to Microsoft Azure*, we'd love to hear your thoughts! Scan the QR code below to go straight to the Amazon review page for this book and share your feedback or leave a review on the site that you purchased it from.

`https://packt.link/r/1804616656`

Your review is important to us and the tech community and will help us make sure we're delivering excellent quality content.

Download a free PDF copy of this book

Thanks for purchasing this book!

Do you like to read on the go but are unable to carry your print books everywhere?

Is your eBook purchase not compatible with the device of your choice?

Don't worry, now with every Packt book you get a DRM-free PDF version of that book at no cost.

Read anywhere, any place, on any device. Search, copy, and paste code from your favorite technical books directly into your application.

The perks don't stop there, you can get exclusive access to discounts, newsletters, and great free content in your inbox daily

Follow these simple steps to get the benefits:

1. Scan the QR code or visit the link below

https://packt.link/free-ebook/9781804616659

2. Submit your proof of purchase
3. That's it! We'll send your free PDF and other benefits to your email directly

www.ingramcontent.com/pod-product-compliance
Lightning Source LLC
Chambersburg PA
CBHW060127060326
40690CB00018B/3790